The Composition and Employment of Software Personnel in the U.S. Department of Defense

An Initial Analysis

BONNIE L. TRIEZENBERG, JASON M. WARD, JONATHAN CHAM, DEVON HILL, SEAN ROBSON, JEFF FOURMAN

Prepared for the Office of the Secretary of Defense
Approved for public release; distribution unlimited

RAND NATIONAL DEFENSE RESEARCH INSTITUTE

For more information on this publication, visit www.rand.org/t/RRA520-1

Library of Congress Cataloging-in-Publication Data is available for this publication.
ISBN: 978-1-9774-0641-5

Cover: tippapatt/Adobe Stock.

www.rand.org

Preface

The U.S. Department of Defense (DoD) has experienced persistent challenges with software development across different kinds of acquisition programs. Attempts to address those challenges through workforce initiatives, such as hiring, training, and professional development, have been severely hampered by an inability to identify and characterize the software professionals—presumably numbering in the tens of thousands—who work within DoD's programs. This report, requested in April 2018 by the Deputy Assistant Secretary of Defense for Systems Engineering, Major Program Support,[1] documents our attempts to identify and characterize DoD's software acquisition workforce, the types of software developed within DoD, and the variety of methods DoD programs use when employing that workforce to develop software. The report should be of interest to policymakers who hire, train, or manage software professionals.

This research was sponsored by the Deputy Assistant Secretary of Defense for Systems Engineering, Major Program Support, and conducted within the Forces and Resources Policy Center and the Acquisition and Technology Policy Center of the RAND National Security Research Division (NSRD), which operates the RAND National Defense Research Institute (NDRI), a federally funded research and development center (FFRDC) sponsored by the Office of the Secretary of Defense, the Joint Staff, the Unified Combatant Commands, the Navy, the Marine Corps, the defense agencies, and the defense intelligence enterprise.

For more information on the Forces and Resources Policy Center or the Acquisition and Technology Policy Center, see their respective websites at www.rand.org/nsrd/ndri/centers/frp and www.rand.org/nsrd/ndri/centers/atp or contact the directors (contact information is provided on each website).

---

[1] As of February 1, 2018, Section 901 of the National Defense Authorization Act for Fiscal Year 2017 abolished the roles of the Under Secretary of Defense for Acquisition, Technology and Logistics and subsidiary Assistant Secretaries. As a result, the project sponsor was restructured under the newly established Assistant Secretary of Defense for Research and Engineering.

Contents

Figures

Tables

Summary

For several decades, the U.S. Department of Defense (DoD) has experienced persistent schedule delays and cost overruns in its software development and software acquisition programs. More than 60 percent of such programs pursued from fiscal year (FY) 2014 through FY 2016 faced major risks, according to recent reviews conducted by DoD's leadership.

DoD's Software Workforce Challenge

Because of the critical role that software plays throughout DoD's everyday operations and because software development is so vital to defense acquisition timelines and spending, DoD in recent years has begun to focus significant managerial attention on its uniformed and nonuniformed software workers and on their counterparts in private industry. In particular, DoD has sought to understand the composition of the workforce that it tasks with acquiring software, as well as the use of current and emerging software development practices and methods by those workers.

But DoD's efforts have been severely hampered by an inability to identify and characterize the thousands of software professionals working within its acquisition programs. These professional do not fit neatly into existing occupational groupings and are spread across the department's existing information technology, program management, engineering, system engineering, test and evaluation, and logistics workforce.

Our Role in Addressing DoD's Challenge

To overcome this workforce knowledge gap, DoD in 2017 approached the RAND Corporation's National Defense Research Institute (NDRI) to assess personnel proficiencies in relevant occupational areas. The goal was to help the department determine better ways to coordinate, define, and manage competencies so that it could more effectively oversee a variety of software functions across the organization. By developing competency models, the project would improve the software acquisition workforce's ability to rapidly and reliably deliver complex software-dependent capabilities.

NDRI conducted its research in two tranches. In the first, the research team developed a model that distinguished 48 competencies related to software acquisition. They also reviewed nearly 400 software training and education courses offered by DoD to identify potential knowledge and preparation gaps among those professional competencies. Ultimately, however, they found that DoD does not have an established system to identify or track employees who perform software functions, nor does it have an accepted set of government job titles or occupations for software professionals. Using those findings, the RAND team recommended that DoD identify with greater specificity who is in the software acquisition workforce.

How Did We Pursue the Current Project?

In the second tranche of the research, which is the focus of this report, NDRI pursued follow-on efforts designed to provide defense leaders with an understanding of (1) the composition of the workforce tasked with acquiring software, (2) the types of software DoD is acquiring, and (3) DoD's use of current and emerging practices and methods for both software development and workforce development and employment.[2]

Table S.1 lists the top-line areas of inquiry and motivating questions that we sought to address in this second tranche of research.

To answer these questions, the research team designed an analytical approach with a data call at its core. The team worked with Country Intelligence Group and the Deputy Assistant Secretary of Defense for Systems Engineering, Major Program Support, to create and conduct a survey of service and selected agencies that would produce data providing a holistic view of software staffing across DoD's major programs.

Although this data call was intended to touch some 50,000 employees throughout the department, it did not occur as designed. As discussed in the next section, the research team encountered several barriers that ultimately proved unsurmountable within the study's resource constraints. Nevertheless, the research team was able to obtain a snapshot of some 2,000 software acquisition personnel employed at a handful of software centers across DoD. That snapshot serves as the basis for the second research tranche's analysis. This snapshot provides sufficient data to (1) identify focal areas for future DoD software competency development initiatives and (2) make initial observations regarding the experience and pay of software professionals working in DoD in comparison with the commercial software industry. Unfortunately, it does not provide the information needed to explore how the dynamics of age, experience, and

---

[2]   The term *workforce employment* is used to denote both the type of software that DoD's workforce is employed to produce and the methods used to organize, train, and equip that workforce.

Table S.1
Research Areas of Inquiry and Questions

Area	Questions
Foundational issues	• What is the composition of DoD's uniformed and nonuniformed software acquisition workforce (by skill set, occupational series, etc.)? • What are the factors associated with successful software acquisition programs? Identify factors that are associated with software acquisition workforce development and employment. • Is the health and stability of a major program correlated to its software acquisition mix?
Workforce employment	• Can we develop a framework to guide DoD program management office software workforce staffing decisions using such factors as program scale, domain, development methodology, complexity, and acquisition approach?
Workforce development	• What changes should DoD make to ensure that the software acquisition workforce can be identified for training, education, and employment purposes? • What software competencies are needed if DoD is to implement modern and emerging software development best practices? • What is the software competency of DoD's software workforce, especially as related to modern and emerging software development best practices? • What changes should DoD implement to existing training and education practices to improve software competency? • What potential barriers should DoD anticipate when implementing modern software practices, and what lessons can it learn from the commercial sector regarding how to mitigate those issues?

pay levels affect the hiring, promotion, and retention of DoD's software professionals. Understanding those dynamics may be essential to designing effective workforce improvement initiatives.

In addition to the snapshot, we conducted structured interviews with selected major programs to obtain a better understanding of the composition and structure of their software workforces, as well as the acquisition approaches they pursue when acquiring software. We also solicited insights from program management offices or equivalent offices regarding competency gaps and opportunities to improve existing software practices.

What Did We Find?

Although the research team gained considerable insight into the distinct challenges of DoD software and software organizations, it was difficult to comprehensively categorize the workforce responsible for that software. The biggest challenges were a lack of data in existing DoD databases and internal DoD organizational issues that limited the team's ability to gather data. Although our research provides a snapshot of DoD's software acquisition workforce, there are significant limitations and possible sources of bias in the underlying data. Our recommendations, which are based on lessons learned from this effort, focus on what is needed to enable future efforts to better understand, train, and employ DoD's software acquisition workforce.

Challenges in Understanding the Composition of DoD's Software Acquisition Workforce

Although the research team's efforts reinforced our belief that a full census data call of DoD's workforce to determine who performs software activities is technically possible and desirable, any subsequent efforts to perform a data call would do well to consider the following barriers that the team encountered:

- fatigue with data calls or surveys across DoD
- inadequate resources within organizations to facilitate data collection
- lack of a perceived benefit to Service Acquisition Executives from data calls and, for some, a belief that software should not be singled out separately from other types of acquisition
- expectations of key stakeholders that low response rates would invalidate the effort
- desires for either (1) a data collection tool other than a spreadsheet, which was felt to be error-prone and inefficient, or (2) direct entry into existing systems, which (although efficient) would not allow for independent vetting of data prior to entering it into authoritative systems
- need for a task memo directing the data call from a level at or higher than Under Secretary of Defense[3]
- pursuit of competing priorities by software supervisors, which results in them having too little time to coordinate data collections in their centers.

Findings Regarding the Composition of DoD's Software Workforce

The research team collected and analyzed a snapshot of about 2,000 DoD software personnel employed at eight of DoD's software development centers and found the following:

- Uniformed personnel, for the most part, do not identify as being part of the software workforce.
- The vast majority (about 80 percent) of DoD software personnel in our sample are in the engineering acquisition career field.[4]

---

[3]   Other DoD-wide data calls for functional matrixed workforces (e.g., the DoD Security Cooperation Workforce) have been successful in achieving high response rates from task memos coordinated at the DoD-component-head level and then signed out at the Under Secretary of Defense level.

[4]   Acquisition career fields within DoD are responsible for human capital functions, including hiring, determining promotions, and providing training and development resources. Being assigned to a career field should not be confused with having a college or university degree in that field. See Chapter Two for a more complete description of the individuals who make up the DoD software workforce.

- About half of DoD's software personnel in our sample have a degree related to software or computer science; more than 90 percent have a degree in science, technology, engineering, or math.
- Most personnel with an educational background in software or computer science spend at least a portion of their time developing software. Less than 3 percent are employed exclusively in management roles.

Distinct Challenges of DoD Software

We collected and analyzed information regarding the software that supports DoD's FY 2018 major programs from different viewpoints, including by number of DoD programs, by program valuation, and by software intensity. Across all metrics, we found the following:

- Most DoD software directly supports the war fight (i.e., software contained in radios, sensors, and weapons, as well as in air, space, and land vehicles).
- Most DoD software is tightly integrated with underlying hardware.
- DoD software's security and safety challenges are far greater than for most commercial software development efforts.

Distinct Challenges of DoD Software Organizations

We augmented our workforce composition data analysis with information gleaned from structured interviews with seven major DoD programs and two investigative groups that had recently concluded in-depth examinations of software issues on DoD programs and found the following:

- Many contributing factors make it difficult to hire software developers into DoD acquisition organizations.
- Once hired, there are substantial barriers to keeping software skill sets current.
- Software project, program, and contract management are key skill sets needed by DoD's software workforce as the department strives to manage a much larger contractor workforce.

What Policy Implications and Recommendations Derive from These Findings?

Our recommendations are grouped into three categories that roughly correspond with Table S.1's research areas: (1) those that DoD should implement immediately to improve its software development process, (2) those that DoD should implement when delivering software training to its software workforce, and (3) those that would allow DoD to better characterize that workforce.

Foundational Issues and Workforce Employment Issues

Our DoD software development process recommendations are as follows:

- It is imperative that DoD develop the competencies and culture needed to incorporate actual warfighters early and often in the development of programs and the underlying software.
- When seeking to reap the benefits of commercial software development practices, DoD must take care not to replicate the security practices of that industry.
- DoD should review current safety standards and guidance to ensure they include best practices for achieving high levels of operational trust in highly automated and/or autonomous DoD software–intensive systems.

Workforce Development and Characterization Issues

Our DoD software workforce development recommendations are as follows:

- DoD software workforce development initiatives should be focused on the best practices of embedded software development.
- Improvement initiatives should focus on engineering acquisition careers, given that most personnel in the DoD software workforce are assigned to that field.

Our DoD software workforce characterization recommendations are as follows:

- Additional efforts are needed to identify and characterize DoD's software workforce. Those efforts must be driven from the highest levels of the DoD hierarchy.
- Future software workforce research teams should include one or more government personnel from the software centers as detached members of the project team.
- Once the software workforce has been identified, DoD should conduct analyses to better understand how the dynamics of age, experience, and pay levels affect hiring and retention.
- DoD should conduct an in-depth analysis regarding the employment of software-educated professionals to ensure that biases regarding suitability for assignment into broader roles are not affecting promotions.

Acknowledgments

We first need to thank the leadership of the software centers, who generously provided us with the snapshot of current DoD software personnel that is the centerpiece of this report. Without them, this report might simply have been a tale of lessons learned. The many participants of the Software Acquisition Workforce Data Call working group also deserve our thanks: Although our attempts to establish an acceptable method to conduct that data call were unsuccessful, their advice throughout was professional, frank, and cogent. We also extend our thanks to Jeff Fourman and the team at Country Intelligence Group for their endless patience and professionalism as they pursued multiple avenues to bring the workforce characterization data collection efforts to a successful conclusion. From the sponsor's office, Bernie Reger was invaluable in connecting us and opening doors. At critical points along our journey, we received timely advice and encouragement from Andy Monje, Sean Brady, Philomena Zimmerman, Jeff Boleng, and the Defense Innovation Board's Software Acquisition and Practices team. Special thanks also to Jerry Tarasek and Tom Hickok, who served as our project liaisons to establish connections with the Functional Integrated Product Teams, coordinate meetings, and provide feedback on project tasks. Finally, we thank our reviewers, Julie Cohen from the Software Engineering Institute and Brad Wilson from RAND, for their feedback and recommendations, which greatly improved the overall quality of this report.

Abbreviations

ABMDS	Aegis Ballistic Missile Defense System
ACQ Demo	DoD Civilian Acquisition Workforce Personnel Demonstration Project
ACS	American Community Survey
AFLCMC	Air Force Life Cycle Management Center
AFRL	Air Force Research Laboratory
CCDC	Combat Capabilities Development Command
CIO	chief information officer
DACM	defense acquisition career management
DAU	Defense Acquisition University
DCPDS	Defense Civilian Personnel Data System
DIB	Defense Innovation Board
DMDC	Defense Manpower Data Center
DoD	U.S. Department of Defense
EB	Armament Directorate
EZAS	Avionics Software Engineering and Integration
FIPT	Functional Integrated Product Team
FY	fiscal year
GAO	Government Accountability Office
GS	General Schedule

IOT&E	independent operational test and evaluation
IT	information technology
NAWCAD	Naval Air Warfare Center Aircraft Division
NAWCWD	Naval Air Warfare Center Weapons Division
NDRI	National Defense Research Institute
NSWC	Naval Surface Warfare Center
OSD	Office of the Secretary of Defense
OUSD(A&S)	Office of the Under Secretary of Defense for Acquisition and Sustainment
OUSD(AT&L)	Office of the Under Secretary of Defense for Acquisition, Technology and Logistics
OUSD(R&E)	Office of the Under Secretary of Defense for Research and Engineering
SAR	Selected Acquisition Report
SEIT	system engineering, integration and test
STEM	science, technology, engineering, and math
TW	Test Wing

CHAPTER ONE

Introduction

For decades, the U.S. Congress and other policymakers have been concerned about schedule delays and cost overruns in U.S. Department of Defense (DoD) acquisition programs. Over the years, the magnitude of these extra costs—sometimes running into billions of dollars—grew to be so substantial that the Government Accountability Office (GAO) deemed a significant number of DoD weapon system acquisitions and DoD business system modernizations to be high-risk.

Particularly challenging in this realm are DoD's software development efforts, especially those running across multiple acquisition programs. According to recent reviews conducted for DoD's Deputy Assistant Secretary of Defense for Systems Engineering and DoD's Chief Information Officer (CIO), more than 60 percent of programs pursued from fiscal year (FY) 2014 through FY 2016 faced major risks in software development, with scheduling and integration accounting for more than half of the risks.[1]

Because of the critical role that software plays throughout DoD's everyday operations and because software development is so vital to defense acquisition timelines and spending, DoD has begun to focus significant managerial attention in recent years on its uniformed and nonuniformed software workers and their counterparts in private industry.[2] In particular, the department has sought to understand the composition of the workforce that it tasks with acquiring software and the use of current and emerging software development practices and methods by those workers.

In 2012, the department completed a comprehensive review of information technology (IT) acquisition competencies, which helped guide the department in successfully establishing a cadre of workers in that field.[3] But DoD leaders soon realized that the department also required complementary competencies in a variety of other

---

[1] Exchange with sponsor in April 2017.

[2] Throughout this document, we refer to DoD's *uniformed* and *nonuniformed* software personnel to distinguish them from *industry* software personnel employed by private-sector commercial entities.

[3] The GAO credited DoD for its successful steps in establishing those acquisition competencies. See GAO, *IT Workforce: Key Practices Help Ensure Strong Integrated Program Teams; Selected Departments Need to Assess Skill Gaps*, Washington, D.C., GAO-17-8, November 2016.

occupations—program management, engineering, system engineering, test and evaluation, and logistics—to assure rapid and reliable delivery of complex software-intensive systems.

Initial Study: Assessing Software Competencies

To gain an understanding of this larger constellation of competencies, DoD asked the RAND Corporation's National Defense Research Institute (NDRI) in 2017 to assess workforce proficiencies in relevant occupational areas.[4] The department sought help with determining better ways to coordinate, define, and manage competencies so that it could more effectively oversee a variety of software functions across the organization. By providing an enhanced understanding of necessary technical competencies and improvements to education, training, workforce management, and assessment, the project's ultimate objective is to improve the ability of DoD's software acquisition workforce to rapidly and reliably deliver complex software-dependent capabilities.

The first tranche of this research was published in 2020.[5] In that report, the research team presented a model that distinguished 48 competencies related to software acquisition. The team also reviewed nearly 400 software training and education courses offered by DoD to identify potential knowledge and preparation gaps among those professional competencies. However, the research team found that DoD has

> no established system for identifying or tracking who performs software functions in DoD. That is, there is no accepted government job title or occupational series for software professionals. Until the software acquisition workforce is identified, it is not possible to take advantage of the competency model or the insights gained from this study on potential gaps in training.[6]

As a result of that finding, the RAND team recommended that DoD identify with greater specificity who is in the software acquisition workforce. Such identifica-

---

[4]   *Competencies* consist of an "observable, measurable pattern of knowledge, skills, abilities, behaviors, and other characteristics (KSAOs) needed to perform work roles or occupational functions successfully." They can support a wide variety of talent management initiatives: recruitment and selection, training and development, career development, and proficiency gap assessments, among others. Collections of competencies for specific career fields or functional areas are generally referred to as *competency models*. The Defense Civilian Personnel Advisory Service adapted this definition from DoD Instruction 1400.25, *Civilian Personnel Management*: Vol. 250, *Civilian Strategic Human Capital Planning (SHCP)*, Washington, D.C.: U.S. Department of Defense, June 7, 2016, p. 21.

[5]   Sean Robson, Bonnie L. Triezenberg, Samantha E. DiNicola, Lindsey Polly, John S. Davis II, and Maria C. Lytell, *Software Acquisition Workforce Initiative for the Department of Defense: Initial Competency Development and Preparation for Validation*, Santa Monica, Calif.: RAND Corporation, RR-3145-OSD, 2020.

[6]   Robson et al., 2020, p. xvii.

tion is critical: Without an understanding of the individuals who constitute the workforce, DoD cannot validate the competency model, nor can it use the model to identify competency gaps in the workforce.

Follow-On Study: Identifying the Software Acquisition Workforce

As the initial study drew to a close, DoD asked the RAND team to pursue a second tranche effort consisting of follow-on research designed to provide defense leaders with an understanding of (1) the composition of the workforce tasked with acquiring software, (2) the types of software DoD is acquiring, and (3) DoD's use of current and emerging practices and methods for both software development and workforce development and employment.

Table 1.1 lists the top-line areas of inquiry and questions that the research team sought to address.

To tackle these questions, the research team (building on the first study's findings and recommendations) designed a research approach to determine the composition of DoD's software acquisition workforce—e.g., numbers of software acquisition personnel, type (uniformed or nonuniformed), grade, service, organization or product center, program, location—and metrics regarding the quality of the workforce relative to identified competencies—e.g., as measured by years of relevant software competency experience, education, knowledge, skills, and abilities.

Table 1.1
Research Areas of Inquiry and Questions

Area	Questions
Foundational issues	• What is the composition of DoD's uniformed and nonuniformed software acquisition workforce (by skill set, occupational series, etc.)? • What are the factors associated with successful software acquisition programs? Identify factors that are associated with software acquisition workforce development and employment. • Is the health and stability of a major program correlated to its software acquisition mix?
Workforce employment	• Can we develop a framework to guide DoD program management office software workforce staffing decisions using such factors as program scale, domain, development methodology, complexity, and acquisition approach?
Workforce development	• What changes should DoD make to ensure that the software acquisition workforce can be identified for training, education, and employment purposes? • What software competencies are needed if DoD is to implement modern and emerging software development best practices? • What is the software competency of DoD's software workforce, especially as related to modern and emerging software development best practices? • What changes should DoD implement to existing training and education practices to improve software competency? • What potential barriers should DoD anticipate when implementing modern software practices, and what lessons can it learn from the commercial sector regarding how to mitigate those issues?

At the heart of the research design was a data call in which the RAND team worked with Country Intelligence Group and the Deputy Assistant Secretary of Defense for Systems Engineering, Major Program Support, to survey service and selected agencies to gain a holistic view of software staffing across DoD's major acquisition programs.[7] This data call was designed to touch some 50,000 employees throughout the department. As we discuss later in this report, the planned data call did not occur. We encountered several barriers that ultimately proved unsurmountable within the resource constraints of our study. We were, however, able to obtain a snapshot of about 2,000 software acquisition personnel employed at a handful of software centers across DoD. That snapshot serves as the basis for the analysis documented here. This snapshot provides sufficient data to (1) identify focal areas for future DoD software competency development initiatives and (2) make initial observations regarding the experience and pay of software professionals working in DoD versus the commercial software industry. Unfortunately, it does not provide the information needed to explore how the dynamics of age, experience, and pay levels affect the hiring, promotion, and retention of DoD's software professionals. Understanding those dynamics may be essential to designing effective workforce improvement initiatives.

In addition to collecting the snapshot, the research team conducted structured interviews with selected major programs to obtain a better understanding of the composition and structure of their software workforce, as well as the scale, domain, application type, development methodology, and acquisition approach for the software that the programs are acquiring. The team also solicited program management office or equivalent office insights regarding competency gaps and opportunities to improve existing software practices, paying particular attention to software workforce employment and development practices. Note that although there was some overlap between the software centers from which we collected workforce characterization data and the program offices that participated in our workforce employment interviews, the two efforts were largely independent.

The relationships between the data call, the structured interviews. and the data analyses are shown in Figure 1.1. The figure depicts the process flow of the research; the areas in which RAND, the sponsor (the Deputy Assistant Secretary of Defense for Systems Engineering, Major Program Support), and the vendor (Country Intelligence Group) were involved; and the disposition of the data that the research team collected.

An additional task not depicted in Figure 1.1 was an analysis to characterize the type of software that DoD acquires. Understanding the type of software being acquired is essential if we are to develop recommendations regarding which software competencies may be most critical in training and assessing DoD software acquisition personnel. Furthermore, characterizing the type of software developed and sustained

---

[7]   This survey included both major defense acquisition programs and major automated information system programs.

Figure 1.1
Research Design

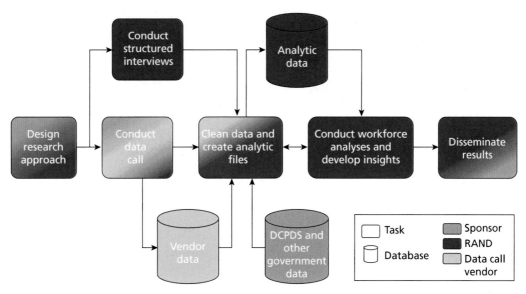

NOTE: DCPDS = Defense Civilian Personnel Data System.

by DoD serves as a "sample frame" to assess whether the data collected for our study are representative of DoD as a whole.[8]

Working Definitions

As discussed in the previous RAND report on DoD's acquisition workforce, DoD has defined and outlined the process by which it designates acquisition positions.[9] Section 1 of the DoD *Defense Acquisition Workforce Program Desk Guide* (2017) defines *acquisition* as

> the conceptualization, initiation, design, development, test, contracting, production, deployment, logistics support (LS), modification, and disposal of weapons

---

[8]   A *sample frame* is a list or other device used to define a researcher's population of interest. The subset (sample) from whom data are collected must preserve the distribution of key characteristics of the full population (the sample frame) to conduct valid research regarding a population. For instance, the most recent census is often used as the sample frame when conducting a poll of U.S. residents. If the distribution of gender and age of respondents to a poll matches that of the census, we say the respondents are a "representative sample for gender and age." In our case, we do not know our population of interest: all DoD personnel performing software activities. Therefore, we used statistics regarding DoD software developed for major programs as a proxy sample frame when assessing the representativeness of our snapshot of DoD software personnel and the validity of the subsequent analysis.

[9]   Robson et al., 2020.

and other systems, supplies, or services (including construction) to satisfy DoD needs, intended for use in, or in support of, military missions.[10]

Positions with more than 50 percent of their duties and responsibilities falling inside this definition are designated and coded as *acquisition workforce* billets.

The previous RAND report also provided a definition of *software acquisition personnel*. Because we started from that definition in this follow-on report, we reproduce it verbatim here, along with relevant explanatory text:

> *Software acquisition personnel* are military, civilian and contractor personnel engaged in the definition, development, deployment, operation, and sustainment of software components and software reliant systems or ecosystems.[11]

> Although this definition is broad, it is designed to differentiate software acquisition personnel from other acquisition professionals such as IT purchasing agents. IT purchasing agents only purchase software (e.g., desktop software); they do not define, develop, deploy, operate, or sustain the software. It is also important to note that some software functions may be performed primarily outside of the acquisition community. Nonetheless, acquisition professionals may need to be at least familiar with these functions to effectively acquire software.[12]

Recall that our initial research design for this follow-on work was a data call of acquisition professionals in selected career fields. As noted in the text quoted above, that approach might miss personnel outside of the acquisition community who nonetheless fit our definition of software acquisition personnel. We avoided this problem in our snapshot by asking the software centers to provide data on all those who define,

---

[10] DoD, *Defense Acquisition Workforce Program Desk Guide*, Washington, D.C., July 20, 2017a, p. 1.

[11] *Development* refers to the processes, procedures, people, material, algorithms, and information required to conceive, specify, design, program, document, test, deliver, and deploy the software aspects of a system. This includes oversight activities required to determine adherence and compliance to contract requirements. *Sustainment* refers to the processes, procedures, people, material, algorithms, and information required to support, maintain, and operate the software aspects of a system. This includes software development, documentation, operations, deployment, security, configuration management, training (users and sustainment personnel), help desk, commercial-off-the-shelf product and license management, and technology refresh. *Software* refers to a collection of data and instructions that executes on a processing unit. Software includes, but is not limited to, applications, scripts, databases, operating systems, device drivers, and firmware. *Software-reliant system* refers to a hardware-software system (such as a radar) that would fail to meet its mission use if the software were to fail or a software system (such as mission planning or intelligence dissemination tools) with its accompanying computing infrastructure and network. *Ecosystem* refers to a set of entities functioning as a unit and interacting with a shared end-user constituency for software and services, together with relationships among them. Ecosystems form when a set of core components (the keystone) are complemented by peripheral components (e.g., apps or services) developed by autonomous entities (i.e., organizationally independent of the core developer) to address specific user needs. Ecosystems are characterized by interoperability and co-innovation enabled through common interfaces and shared knowledge.

[12] Robson et al., 2020, pp. 7–8.

develop, deploy, operate, or sustain software. As a result, we refer to these personnel simply as *DoD software personnel* in this report.[13] As used in this report, DoD software personnel include both nonuniformed and uniformed personnel employed by DoD who engage in software activities for the department.[14]

Organization of This Report

In Chapter Two, we offer some preliminary answers to the basic questions regarding the personnel who make up DoD's software workforce. Although we were unable to gather consistent data across DoD regarding the workforce, we use data collected over the course of our research as a snapshot to provide insight into the mix of uniformed and nonuniformed personnel employed by DoD and their acquisition career fields, educational backgrounds, experience, and pay scales. In Chapter Three, we turn our attention to characterizing the type of software that DoD's workforce is charged with acquiring and reflect on the implications of our findings. As noted earlier, these software characterizations not only form the basis for understanding which software competencies may be most critical in training and assessing DoD software personnel but also serve as a sample frame to assess whether the data collected for our study are representative of DoD as a whole. In Chapter Four, we summarize what we learned from our interviews regarding how software personnel are employed within programs and regarding the barriers program managers and software leadership are encountering as they strive to modernize their workforce and development practices. Because we continue to believe that DoD will need to make a concerted effort to track and train its software workforce, we detail in Chapter Five our extensive efforts to implement a DoD-wide data call of that workforce. We also provide recommendations that should prove useful if others are to avoid the pitfalls we encountered. Finally, we provide a summary of our study in Chapter Six.

In Appendix A, we discuss our data collection methodology and sources of potential bias for both the workforce characterization and workforce employment portions of our study. In Appendix B, we discuss the data analysis methodology used to characterize DoD's software workforce and to compare DoD's workforce with the software industry more generally. Some of our analyses did not rise to the level of accuracy that we believe is required to provide actionable insight, given the substantial assumptions

---

[13] Our decision to drop the term *acquisition* when characterizing DoD's workforce who perform software activities was informed by recent moves within DoD to officially recognize that software does not fit neatly into acquisition versus sustainment phases. See Defense Innovation Board, *Software is Never Done: Refactoring the Acquisition Code for Competitive Advantage*, May 3, 2019.

[14] In this report, we elected not to use the word *civilian* when referring to the subset of DoD personnel who are not in uniform. This is to avoid confusion with those civilians who are employed by commercial software firms. We use the term *software industry personnel* to refer to this latter group.

we made when constructing comparisons of DoD and industry. For those analyses that do, we report the results in Chapter Two. For those that do not, results are shown in Appendix B for completeness and to serve as an example of the type of future analyses that should be done if DoD is to develop a more complete understanding of the dynamics that drive recruitment and retention of this workforce. Appendix C contains further visualizations and service-level breakdowns of the DoD workforce characterization data used in our research.

CHAPTER TWO

Snapshot of DoD Software Personnel

In this chapter, we provide a summary of the policy-relevant observations from our analysis of the snapshot of about 2,000 DoD software personnel obtained from our workforce data collection efforts. Action officers from eight of DoD's software centers identified personnel engaged in software activities at their centers and then collected and provided to us a limited set of data about that workforce. For perspective, this chapter compares the DoD's workforce sample with personnel in the general U.S. software industry for selected characteristics.[1]

Because of difficulties with collecting data regarding DoD's workforce, the sample of software personnel obtained for this analysis from the DoD software centers might not be representative of DoD as a whole. However, when compared with the relative software value of major DoD programs, the sample is reasonably representative of the service organizations that procure DoD systems, albeit slightly overweighted toward Army programs (Table 2.1). Details regarding our computation of the percentage of software value, derived by analyzing FY 2018 major programs, are provided in Chap-

Table 2.1
Affiliation of Sampled Workforce Versus the FY 2018 Major Program Software Value

Affiliation	Percentage in Sample	Percentage of FY 2018 Major Program Software Value
Army Program	32	16
Navy Program	54	65
Air Force Program	12	18
Joint Program	2	2

NOTE: The highlighted row is indicative of the overweight of Army programs in our sample. Although we cannot claim extensibility across DoD, the sample is reasonably well balanced across the DoD service organizations, and we believe that basic observations gleaned from our analyses are relevant.

---

[1] General U.S. industry workforce data were derived from the American Community Survey (ACS) microdata from 2016 to 2018. More-detailed information regarding the DoD workforce data collection process can be found in Appendix A.

ter Three.[2] Software personnel data were collected in late 2019 and early 2020. Details regarding our data collection methodology for this task are provided in Appendix A, and details of the analyses are provided in Appendix B.

Mix of Uniformed and Nonuniformed DoD Software Personnel

An interesting observation regarding our snapshot is that, despite a recent push to develop soldier coders, individuals who were categorized as software personnel in our sample are 99 percent nonuniformed personnel; just 1 percent are uniformed enlisted personnel or officers.[3] Unfortunately, we did not have this data early enough in our research to follow up during our interviews. Therefore, we do not know if the relative absence of uniformed personnel within our sample is because (1) few uniformed professionals engage in software activities, (2) there was a reluctance to categorize uniformed professionals as part of the software workforce, or (3) these soldier coders exist within DoD but are employed outside the software centers. As we will discuss elsewhere in this report, engaging actual warfighters in the software development process is highly leveraged. Certainly, improving our understanding of the mix of uniformed and nonuniformed personnel engaged in software activities is an area for which later efforts to characterize DoD's software workforce will want to follow up.

Acquisition Career Field of DoD Software Personnel

Another policy-relevant observation is that the vast majority (about 80 percent) of personnel engaged in software development at the centers are in the engineering acquisition career field. Acquisition career fields within DoD are responsible for many human capital functions, including hiring, determining promotions, and providing training and development. Functional Integrated Product Teams (FIPTs) within DoD are aligned to these career fields and advocate for career-specific initiatives to improve training and support career development. Although the IT FIPT has been particularly active over the last several years in supporting software engineering training and

---

[2]   The percentage of software value acquired by each service reflects our characterization of software acquired for DoD FY 2018 major acquisition programs (discussed in Chapter Three). Therefore, representativeness is with respect to those major software development efforts. If later workforce data collection efforts find that DoD spends a significant portion of its software efforts outside of major programs, observations made in this chapter may need to be revised.

[3]   The push for military coders gained momentum in 2018, coincident with our study. See James Long, "Army of Coders: Training the Force for the Multi-Domain Fight," Modern War Institute, December 21, 2018. It is possible that our snapshot was simply taken too soon to see the results of those efforts. This is an area for which the ability to code existing DoD personnel databases to indicate those personnel performing software activities would allow us to monitor trends in the mix of uniformed and nonuniformed DoD personnel.

career development initiatives, we found just 1 percent of personnel engaged in software activities at the centers were associated with the IT career field. Another 9 percent of personnel engaged in software activities are in the System Engineering or Integration and Test career fields, and another 9 percent are in the Program Management or Science and Technology career fields. This leads us to the following recommendation: *Software competency improvement initiatives should be focused on personnel in the engineering career field.* By focusing first on the engineering career field, DoD will engage the majority of personnel performing software activities. The proportion of individuals engaged in software activities by career field is shown in Figure 2.1.

Educational Background of DoD Software Personnel

A third observation from our analysis is that just over half of those engaged in software activities have an educational background in a software-related or computer science–related field. The second largest group of personnel engaged in software activities (about 30 percent) are those with an aerospace or electrical engineering degree, which is not surprising, given the software-intensive nature of avionics and signal-processing functions that make up aerospace and electrical systems. Close to 10 percent of the workforce have another type of engineering degree, 4 percent have degrees in other science, technology, engineering, and math (STEM) fields, and only 4 percent have degrees in a non-STEM field. Taken in total, the educational backgrounds of DoD

Figure 2.1
Career Field of Individuals Engaged in Software Activities

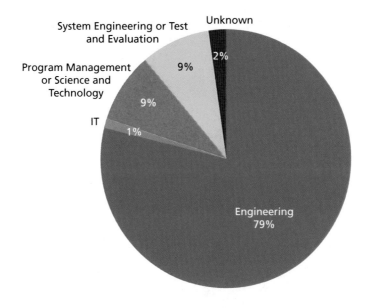

software personnel reveal a highly technical workforce.[4] Although the distribution of educational backgrounds we found within our snapshot (shown in Figure 2.2) may be biased by the limited number and type of centers we obtained data from, it aligns with the experience of many of those in the program offices we interviewed.[5]

We also examined how personnel with a software-related or computer science–related background are employed at the centers.[6] As shown in Figure 2.3, we did not have sufficient data to characterize the roles of about 30 percent of the workforce with an educational background in software. However, we were able to ascertain that at least half of personnel with a software background spend a portion of their time developing software. A much smaller percentage are employed in management or system engineering, integration and test (SEIT) roles.

These observations led us to explore the educational background of personnel engaged in software development by their role. As shown in Figure 2.4, the developer role is dominated by those with a degree in software or computer science. As roles become less directly involved with the actual production of software, the distribution of educational backgrounds becomes less specialized. In fact, of those for whom management is their only software development–related activity, educational background is distributed over the range of categories we used in our analysis. Even in this case, however, technical degrees dominate.

The relative lack of software-educated professionals in roles focused on management of software should be investigated further. We are concerned that it may be an indicator that systematic biases within DoD are limiting opportunities for these individuals. In light of our observation from program interviews that chief engineers often do not have a background in software development (see Chapter Three), we are

---

[4]   For perspective, a 2019 study of the overall DoD nonuniformed workforce showed that the overall portion with a background in STEM fields is around 10 percent (Spencer T. Brien, *Attrition Among the DoD Civilian Workforce*, Monterey, Calif.: Naval Postgraduate School, NPS-HR-20-004, October 2019). An earlier DoD acquisition workforce study found that the acquisition workforce, in general, has higher educational levels than DoD overall but does not give a breakdown of the types of degrees obtained (Susan M. Gates, Edward G. Keating, Adria D. Jewell, Lindsay Daugherty, Bryan Tysinger, Albert A. Robbert, and Ralph Masi, *The Defense Acquisition Workforce: An Analysis of Personnel Trends Relevant to Policy, 1993–2006*, Santa Monica, Calif.: RAND Corporation, TR-572-OSD, 2008, p. 13, Figure 3.4).

[5]   A lack of educational background in software should not be interpreted as a lack of skill in software. Given the relatively recent emergence of computer science and software engineering as a field of study, older personnel (in particular) may have gained software skills outside a formal educational setting.

[6]   For this analysis, we categorized individuals into three software roles (developer, SEIT, and manager) according to their answers regarding the type of software activity performed. This classification was by process of elimination: Personnel who performed software architecture, design, implementation, or unit test activities were classified as developers; personnel who did not perform those activities but did perform SEIT activities (such as requirements analysis, integration, and test) were classified as SEIT; and the remaining individuals who only perform planning and oversight tasks were classified as managers. Therefore, a manager is not a position but a role: i.e., a hands-on manager who performs all software activities, including design and implementation, is characterized as a developer using this classification scheme.

Figure 2.2
Educational Background of Individuals Engaged in Software Activities

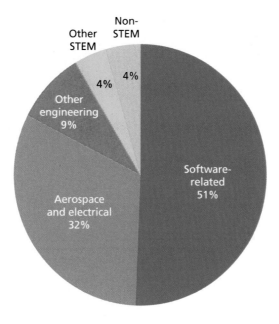

Figure 2.3
Employment of Individuals with an Educational Background in Software or Computer Science

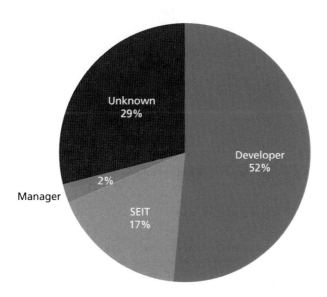

Figure 2.4
Educational Background by Software Role

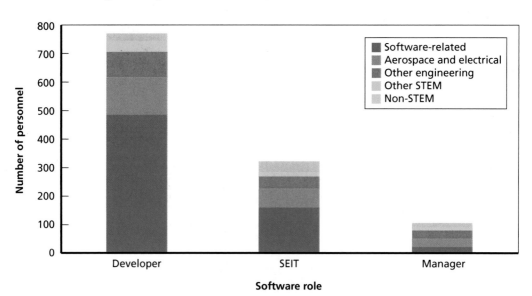

concerned that personnel with software skill sets may be at risk of being overlooked in assignment to broader system and/or management roles. We are not asserting that we found this to be the case, simply that we found enough evidence to warrant further investigation.[7] *We recommend DoD conduct an in-depth analysis regarding the employment of software-educated professionals to ensure that biases regarding their suitability for assignment into broader roles are not affecting promotions.*

Experience and Pay-Scale Distributions of DoD Software Personnel Versus Commercial Industry

We also performed analyses to compare DoD's workforce to the software industry more generally. Details of the analyses and the construction of the data sets and variables therein are provided in Appendix B.[8] From these analyses, we generated our third

---

[7]   Although a small subset of the centers provided us job titles for their personnel, these data were not sufficient to allow us to explore this hypothesis further. Ideally, an analysis would look at the software competency of program management and chief engineer positions for software-intensive programs. Because we do not yet have a way to measure software competency, educational background could be used as a reasonable proxy.

[8]   Appendix B also contains additional analyses that illustrate the types of analyses DoD should do to characterize its workforce if it is to develop the understanding needed to recruit and retain software professionals. We chose not to include these analyses in the main body of this report, however, since our data set required that we make extensive assumptions when interpreting that data. The biases introduced in making those assumptions may render the results invalid, and we caution readers not to rely on them when designing workforce initiatives.

relevant finding: *Any analysis that focuses on the "average" age, experience, or pay scale of software personnel in DoD will miss significant dynamics regarding that workforce.*

This finding is based on our observation that software personnel in our DoD sample either tend to have recently entered the field or tend to have long experience in it.[9] Because we did not collect demographic data on our sample, we cannot say whether this reflects an age distribution (e.g., most are young or old with very few midcareer professionals) or whether recent demand spikes for software professionals have caused more personnel of all ages to recently enter the field. To explore whether this is an age-based phenomenon that might also appear in the commercial software industry, we used "years of age minus years of education" as a proxy for career experience for individuals in the ACS who indicated they were employed in software-related industries.[10] We expect to find the broad patterns comparable unless many people in the commercial software industry workforce move into that sector after working in other industries in a systematically different fashion than DoD software personnel. Comparisons with the U.S. total workforce and the U.S. software industry workforce are shown in Figures 2.5 and 2.6.

Given the significant differences between the profiles of the DoD software and industry profiles, we make the following observations:

- DoD's software workforce experience profile is not likely to be age-based.
- It is more likely that the experience profile is being driven by (1) factors that are inherent to a young industry (such as software), (2) distinct DoD hiring and retention policies, and/or (3) unusual aspects of our DoD workforce snapshot.

The above comparison of experience profiles between DoD and industry may indicate that salary is influencing the hiring and retention of DoD software personnel.[11] Our data allow us to conduct a limited analysis of salary differentials between DoD and industry. The results, shown in Figure 2.7, are segregated by educational background.

For those with a STEM degree, DoD salaries start at a comparable level with industry pay but do not keep pace as experience increases. In fact, for those with a STEM degree and more than ten years of experience, we see estimated salary gaps of $25,000 to $45,000 per year relative to private industry compensation. A different

---

[9]   Data in Figure 2.5 plot responses to the "years of software experience" for those who indicated they were engaged in software activities. Where that field was blank, we used "years in current position."

[10]   The industry we restrict the ACS sample to is "computer systems design and related services." To make the distribution of education in the ACS data more comparable to the DoD data, we drop those without at least an associate's degree.

[11]   Another driver of the significantly different experience profile seen within DoD could be the lingering effects of DoD's 1990s hiring policies. Understanding the dynamics of hiring and retention policies *over time* requires that DoD find a way to identify software personnel within existing personnel databases.

Figure 2.5
Workforce Experience Profiles in DoD Sample

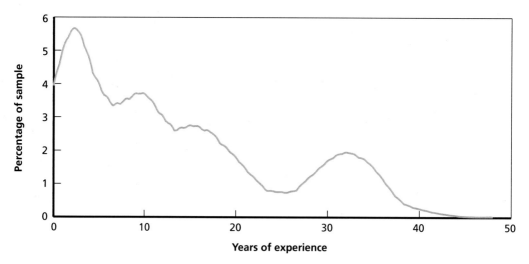

SOURCE: RAND analysis of collected DoD software personnel data and of U.S. industry workforce data derived from ACS microdata from 2016 to 2018.

Figure 2.6
Workforce Experience Profiles in Industry Sample

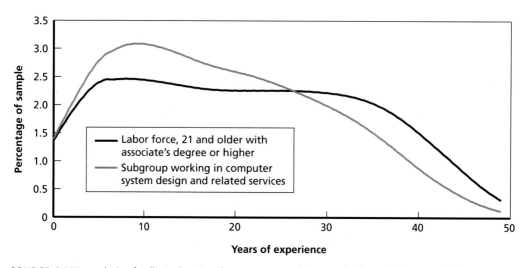

SOURCE: RAND analysis of collected DoD software personnel data and of U.S. industry workforce data derived from ACS microdata from 2016 to 2018.

Figure 2.7
Salary Differentials (DoD Minus Industry) by Experience and Degree Category

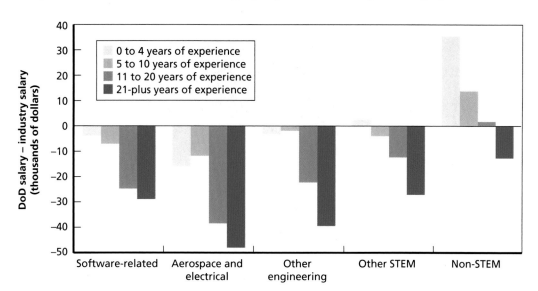

SOURCE: RAND analysis of collected DoD software personnel data and of U.S. industry workforce data derived from ACS microdata from 2016 to 2018.

pattern is observed for non-STEM degrees. In this case, there appears to be a premium in initial DoD pay of around $25,000 per year. Those with high levels of experience do see a salary gap, but it is lower than that seen by professionals with STEM degrees (i.e., around $15,000 per year).

Note that for the differential in Figure 2.7 to be valid, we need to assume that age in our industry data is a reasonable proxy for years of experience in the DoD sample. As we discuss in more depth in Appendix B, this assumption may not be valid, given the relative youth of the software profession. If experience is a stronger determinate of salary than age in the industry sample, then the higher salaries in industry that we found at higher years of experience are understated. If age is more important than experience in determining DoD salaries, then the comparisons may be valid.[12]

Because we observed reversed pay differentials for personnel with STEM degrees versus personnel with non-STEM degrees, we took a closer look at how educational background affects the experience profiles of DoD. Plotting years of experience by educational background of our DoD sample (Figure 2.8) shows that late-career software professionals within DoD primarily have "other engineering" or "non-STEM" degrees. In combination with our observation that personnel with non-STEM degrees

---

[12] Note that both DoD and industry salaries ideally are merit-based. Although neither age nor years of experience necessarily imply greater competency, it is true that years of experience is a closer approximate of competency than age (assuming we learn from our experience). In actuality, both the industry and DoD samples show a strong correlation of salary to years of experience (for DoD) and to age (for industry). See Appendix B.

Figure 2.8
DoD Software Personnel Experience Profiles, by Educational Background

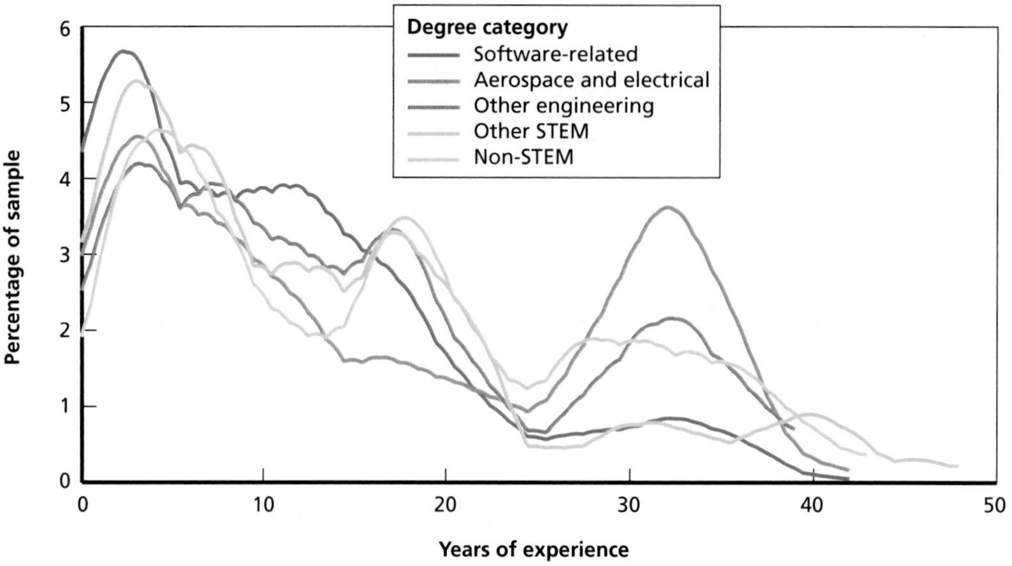

SOURCE: RAND analysis of collected DoD software personnel data.

may have higher salaries in DoD versus industry, we believe there is sufficient evidence to state that pay differentials for those educated in STEM fields may be affecting DoD's ability to retain and/or recruit experienced software professionals.[13]

To explore how these differences in pay and experience distribution may be playing out in promotion and retention, we conducted additional analyses to compare pay by census occupation group and years of experience for those at higher pay grades versus the workforce as a whole. As documented in Appendix B, these analyses are flawed because the variables of interest are not present within our data set and creating them required that we make several assumptions about how to interpret the data we have. For this reason, they are not included in the main body of this report. Although we are unwilling to draw findings or even offer observations from these analyses, we elected to document them in Appendix B to give our readers perspective on the types of questions that need to be answered if DoD is to design meaningful initiatives to improve the competency of DoD's software acquisition workforce.

---

[13] Because the number of software professionals in all educational backgrounds is at its low point for those with 25 years of experience but improves with additional years of experience for those with non-STEM and "other engineering" degrees, we suspect some of these professionals are entering or reentering the DoD workforce after retirement. Although this phenomenon may just be a feature of the software centers that provided our sample, it suggests that an exploration of retirement policies on workforce recruitment and retention could yield significant insight. See Appendix B for alternate explanations of the DoD software workforce experience profile.

Summary of DoD Software Workforce Characterization

Key findings from our analysis of the snapshot of personnel who participate in software development activities at DoD software centers include the following:

- Uniformed personnel, for the most part, did not identify as being part of the software workforce.
- The vast majority (about 80 percent) of DoD software personnel in our sample are in the engineering acquisition career field.
- About half of DoD's software personnel in our sample have a degree related to software or computer science; more than 90 percent have a STEM degree.
- Most personnel with an educational background in software or computer science spend at least a portion of their time developing software. Less than 3 percent are employed in management-only roles.

From our efforts to understand the dynamics of this workforce, we noted the following observations and concerns:

- Any analysis that focuses on the "average" age, experience, or pay scale of software personnel in DoD will miss significant dynamics regarding that workforce.
- Pay differentials between DoD and industry for those educated in STEM fields may be affecting DoD's ability to retain and/or recruit experienced software professionals.

From these findings and observations, we make four recommendations:

- Improvement initiatives should be focused on the engineering acquisition career field, given that most personnel in the DoD software workforce are assigned to that career field.
- Additional efforts are needed to identify and characterize DoD's software workforce.
- Once the software workforce has been identified, DoD should conduct analyses to better understand how the dynamics of age, experience, and pay levels affect hiring and retention.
- DoD should conduct an in-depth analysis regarding the employment of software-educated professionals to ensure that biases regarding suitability for assignment into broader roles are not affecting promotions.

CHAPTER THREE

Characterizing DoD Software Acquisitions

In this chapter, we turn our attention to characterizing the type of software that DoD's workforce is charged with acquiring and reflect on the implications of our findings. Understanding the type of software being developed is essential if we are to develop recommendations regarding which software competencies may be most critical when training and assessing DoD software acquisition personnel. Furthermore, characterizing the type of software developed and sustained by DoD serves as a sample frame to assess whether the data collected for our study are representative of DoD as a whole. We limited our analysis to DoD's FY 2018 major programs.[1] These programs include not just "new" software development efforts but also programs in which software is being added to improve the functionality or extend the use of existing systems. In this sense, the programs could be termed to cover both the acquisition and sustainment phases of software development.[2] However, if later workforce data collection efforts find that DoD spends a significant portion of its software effort outside of major programs, observations made in this report may need to be revised.

Although characterization using a simple metric that represents the percentage of total programs may be of interest, the FY 2018 major programs vary greatly in terms of their dependency on software and in terms of total value. Therefore, we set out to construct a more relevant metric for our analysis. In pursuit of that goal, we used independent subject-matter experts from RAND and within the sponsoring organization to categorize each program by type and by software intensity.[3] We also assigned a program value to each effort, derived from program office estimates for research, development, test, and evaluation (RDT&E), as recorded in Selected Acquisition Reports

---

[1] DoD, "Major Defense Acquisition Programs (MDAP) and Major Automated Information Systems (MAIS) List," April 1, 2018.

[2] In the past year, DoD acquisition processes for software have been modified to reflect that software does not experience distinct acquisition and sustainment phases (Defense Innovation Board, 2019).

[3] Software intensity was scored on a scale of 0 to 5, with 0 representing programs that do not include software development and 5 representing programs whose primary product is a software system. A system that is equally dependent on hardware and software is a 3 on this scale. Note that the scale captures our experts' evaluation of the system's *dependency* on software, not on the cost of developing the software.

(SARs). [4] The total program value of these major acquisition programs for FY 2018 is in excess of $800 billion. In the tables in this chapter, we report metrics for each categorization of DoD software by (1) percentage of FY 2018 programs, (2) percentage of total program value, and (3) percentage of what we term *software value*: the program value when weighted by software intensity. Note that the percentage of software value thus represents a relative measure of DoD development that is "at-risk" if DoD's workforce does not have the required competencies needed to develop that category of software. For this reason, we use the percentage of software value as the basis for our recommendations. Although our analysis may not be generalizable to the whole of DoD, we believe it provides significant insight into major program risk driven by software.

Warfighter Support Software Development Dominates DoD Acquisitions

A guiding principle of modern software development is that the voice of the customer must be central to the development effort. Therefore, our first categorization of the FY 2018 DoD major programs is focused on who represents the voice of the customer for that system. In our analysis, we categorize each program by what we term the *supported entity*: warfighters (those on the battlefield), commanders and planners at headquarters or operations centers, and those who provide business support and logistics to the defense enterprise. Additionally, there are a handful of programs that provide integrative support to all of these entities, which we categorize as *weapon systems*. The overall percentages are shown in Table 3.1.

The largest category, warfighter support, includes programs that produce radios, sensors, and the air, space, and land vehicles used in warfighting. These are the systems that personnel on the battlefield most often interact with, and they represent the largest "supported entity percentage" by each of our metrics. The second-largest category by percentage of software value is commander support, which includes battle management and command post systems. Weapon systems and the business and logistics systems categorized as supporting the defense enterprise each represent only 2 percent of software value.

---

[4]   Total program value includes the cost of developing both hardware and software. Although estimates in SARs are rarely accurate, we believe this analysis adequately portrays the *relative value* of the systems that are "at-risk" if DoD does not have the expertise in a particular software development domain. A still relevant report on the inadequacies of SAR data is Paul G. Hough, *Pitfalls in Calculating Cost Growth from Selected Acquisition Reports*, Santa Monica, Calif.: RAND Corporation, N-3136-AF, 1992. As recently as 2013, a RAND study found that the accuracy of SAR program office estimates ranged from –10 percent to +275 percent of total program value, with a typical program inaccuracy of +75 to +115 percent (Robert S. Leonard and Akilah Wallace, *Air Force Major Defense Acquisition Program Cost Growth Is Driven by Three Space Programs and the F-35A; Fiscal Year 2013 President's Budget Selected Acquisition Reports*, Santa Monica, Calif.: RAND Corporation, RR-477-AF, 2014).

Table 3.1
FY 2018 Major Programs, by Supported Entity

	Weapon Systems	Warfighter Support	Commander Support	Defense Enterprise Support
Count	23	82	23	11
Percentage of programs	17%	59%	17%	8%
Percentage of total program value	8%	86%	6%	1%
Percentage of software value	2%	89%	7%	2%

NOTE: Software value is the program value weighted by software intensity. Software intensity is scored on a scale of 0 to 5 to denote the system's *dependency* on software, not the cost of development. Values are rounded to the nearest whole percentage and may not sum exactly to 100 percent.

Although warfighter support programs represent just 59 percent of the total programs analyzed, they represent 86 percent of total program value and 89 percent when program value is weighted by software intensity. Therefore, improving the competency of warfighters to provide an effective voice of the customer to software development teams is highly leveraged. Unless real battlefield lessons are incorporated into the operation of these systems, the underlying software is unlikely to fully meet warfighter needs. Yet taking warfighters off the battlefield to support software development is rarely done, even in relatively benign operational environments, such as spacecraft operations.[5] This observation leads us to our first recommendation: *It is imperative that DoD develop the competencies and culture needed to incorporate actual warfighters early and often in the development of programs and the underlying software.* As we will discuss in Chapter Four, program managers report both cultural and organizational barriers when attempting to meet this goal.

Most Software Developed by Major DoD Programs Is Tightly Integrated with Underlying Hardware

Our second categorization of the FY 2018 major programs was between the software that is embedded in physical systems (e.g., radars, aircraft, satellites, radios) and that which resides on enterprise computing infrastructure (e.g., planning or battle management software on laptops, desktop computers, and servers or in the cloud).[6] These

---

[5] In a 2019 review of space programs, the GAO found that all four programs it analyzed struggled to effectively engage end users during development (GAO, *DoD Space Acquisitions, Including Users Early and Often in Software Development Could Benefit Program*s, Washington, D.C., GAO-19-136, March 2019).

[6] Our definition of *embedded software* includes any software that is tightly integrated with its underlying hardware, not only software that is embedded in a custom chip. A defining characteristic of embedded systems is the

two classes of software require very different skill sets, both during the initial development of the software and when later improvements are made. Of the 139 systems we examined, just 27 (19 percent) fell into the enterprise computing category, and those programs represented only 3 percent of total program value and 5 percent of the total software value. Although it is true that sometimes 20 percent of software can result in 80 percent of overall risk, it does not appear from this simple analysis that improving DoD competencies in enterprise computing is highly leveraged. Scarce resources may be better focused on embedded computing.

In an attempt to gain further insight into the nature of DoD's embedded systems, we segregated out vehicles (e.g., air, naval, and land combat vehicles) and communications systems (e.g., radios, satellite terminals, communication satellites) from embedded systems more generally. Table 3.2 shows the breakdown of enterprise computing versus different embedded systems by total program value and software value (i.e., program value weighted by software intensity). Vehicle systems, with the requirement

Table 3.2
FY 2018 Major Programs by Software Type

	Enterprise	Embedded (Vehicles)	Embedded (Communications)	Embedded (General)
Count	27	49	20	44
Percentage of programs	19%	35%	14%	32%
Percentage of total program value	3%	80%	4%	8%
Percentage of software value	5%	81%	12%	7%

NOTE: Software value is the program value weighted by software intensity. Software intensity is scored on a scale of 0 to 5 to denote the system's dependency on software, not the cost of development. Values in table are rounded to the nearest whole percentage and may not sum exactly to 100 percent.

requirement to operate in real time within constrained computing resources. As stated in Techopedia's definition,

Embedded software can be . . . quite complex such as the software running all of the electronic components of a modern smart car, complete with climate controls, automatic cruising and collision sensing, as well as control navigations. Complex embedded software can also be found in aircraft avionics systems, in very complex fly-by-wire systems used in fighter planes and even in missile guidance systems (Techopedia, "Embedded Software," webpage, undated).

Our definition of embedded systems is perhaps closest to what the National Institute of Standards and Technology refers to as *cyber-physical systems*. The organization defines these systems as "co-engineered interacting networks of physical and computational components. See National Institute of Standards and Technology, "Cyber-Physical Systems," webpage, undated. The GAO notes that DoD also sometimes uses the term *platform information technology* to refer to these systems: "software, that is physically part of, dedicated to, or *essential in real time to the mission performance of special purpose systems*" (emphasis added; GAO, *Weapon Systems Cybersecurity: DoD Just Beginning to Grapple with Scale of Vulnerabilities*, Washington, D.C., GAO-19-128, October 2018).

In our work developing DoD software acquisition competencies, we originally referred to these systems as *cyber-physical* but adopted the word *embedded* on the advice of our reviewers.

to integrate multiple software systems into a functional whole, represent just 35 percent of the programs but dominate DoD's major acquisitions under both valuation methods (i.e., both their program value and software value are greater than 80 percent of the total).

To understand whether vehicle systems also dominate troubled DoD programs, we categorized Nunn-McCurdy breaches for calendar years 2007–2015 and discovered that 15 of the 29 programs (52 percent) that experienced breaches in those years were vehicle systems.[7] This finding suggests that *integration of vehicle systems is one of the most highly leveraged competencies required of DoD software professionals.*

In our analysis of Nunn-McCurdy breaches, we also noted that communication systems (satellites, radios, and gateways) have had an outsize percentage of breaches, suggesting that their complexity has been historically underestimated. These embedded systems face significant performance constraints in terms of latency, throughput, and memory usage. A failure to recognize the need to balance those performance constraints against other program quality attributes (extensibility, interoperability, maintainability, etc.)—often described as a failure to design for scalability—has been a root cause of severe programmatic issues on recent DoD software programs.[8]

These observations lead us to our second recommendation: *DoD software workforce development initiatives should be focused on the best practices of embedded software development.* Improving DoD's competency in developing real-time embedded software is essential if we are to field usable radars, radios, aircraft, ships, and satellites. This is not to say that improving competencies in cloud and enterprise software development best practices cannot be leveraged to improve the productivity of those who

---

[7] A Nunn-McCurdy breach occurs when a program's cost exceeds thresholds set as a percentage of original and current baseline cost estimates. We analyzed the list of Nunn-McCurdy breaches from Mosche Schwartz and Charles V. O'Connor, *The Nunn-McCurdy Act: Background, Analysis, and Issues for Congress*, Washington, D.C.: Congressional Research Service, R41293, May 12, 2016.

[8] For example, in a report dissecting failures to produce the Joint Tactical Radio System (JTRS) in a timely fashion, the authors write,

> The JTRS program's failure to define the specific limitations of the available technology, and, instead, rely on a "responsive" and "flexible" architecture, inculcated the belief that difficult technical problems could be addressed at some other point in the development process (Jacques S. Gansler, William Lucyshyn, and John Rigilano, *The Joint Tactical Radio System: Lessons Learned and the Way Forward*, College Park, Md.: University of Maryland Center for Public Policy and Private Enterprise, revised February 2012, p. viii).

Even enterprise and cloud computing systems can experience computing performance issues if not designed appropriately for *scalability* (a system's ability to maintain performance as the demand for services increases). In a 2012 report, DoD's Developmental Test and Evaluation organization cited scalability as a critical deficiency in the Joint Space Operations Center Mission System software—software that we classify as enterprise computing (DoD, *Developmental Test and Evaluation: FY 2012 Annual Report*, Washington, D.C., March 2013). Scalability issues have also surfaced with DoD's health records systems: Evaluators found unacceptable delays in the system in response to user inputs (Office of the Secretary of Defense, *Military Healthcare System (MHS) GENESIS Operational Test and Evaluation (IOT&E) Report*, Washington, D.C., April 30, 2018).

develop embedded software, only that it will not directly improve the quality of the embedded programs that make up the vast majority of DoD software development.[9]

Security and Safety Challenges Are Far Greater Than Most Commercial Development Efforts

The third and fourth categorizations we applied to the FY 2018 major programs were security criticality and safety criticality. For our definition of *security-critical software*, we adopted the definition proposed by Gutgarts and Temin in a 2010 paper: "Software is security-critical when it is expected to operate orderly in a hostile environment."[10] Unsurprisingly, we found no DoD major programs that were not security critical. Even DoD's enterprise support systems are under increasing attack and must operate through those attacks. Furthermore, several of the FY 2018 major programs are developing the cryptography and key infrastructure that protect U.S. national security interests from hostile intent, perhaps the most security critical of all software within DoD.

Competency in developing security-critical software is in short supply within the software industry. Although a recent report on web application security risks showed the software industry is improving in their security practices, the report still found that nine out of ten web applications are vulnerable to attack, 39 percent allow unauthorized access to applications, and 68 percent are vulnerable to breaches of sensitive data.[11] Microservice architectures increase the attack surface of systems by creating more ingress points and may make those systems increasingly vulnerable to man-in-the-middle attacks.[12] Understanding the security strengths and weaknesses of these architectures is essential as DoD drives for wider adoption of microservices to provide

---

[9]   For a concrete example, understanding how Facebook rolls out new software every few hours is unlikely to change how a developer of radar software—which runs on highly specialized hardware to implement the precise timing needed to generate, receive, and process the radar signal—deploys and delivers the radar software itself. However, at least some of the methods in use at Facebook may be applicable to the deployment and delivery of the radar software *development environment*: a mix of commercial-off-the-shelf and custom software used to design, code, build, configuration manage, track issues, and execute automated checks of the radar software. Improving the efficiency of deploying and delivering the software development environment directly improves the productivity of the software workforce developing the radar. In fact, DoD is making significant investments in standardizing and packaging common software tooling for its software development teams. The principles guiding those investments are outlined in U.S. Department of Defense Chief Information Officer, *DoD Enterprise DevSecOps Reference Design*, Washington, D.C., August 12, 2019.

[10]   Peter B. Gutgarts and Aaron Temin, "Security-Critical Versus Safety-Critical Software," *Proceedings*, 2010 IEEE International Conference on Technologies for Homeland Security (HST), Waltham, Mass., November 8–10, 2010.

[11]   Positive Technologies, *Web Application Vulnerabilities and Threats: Statistics for 2019*, February 13, 2020.

[12]   B. Cameron Gain, "Microservices Security: Probably Not What You Think It Is," *New Stack*, March 26, 2018.

greater composability and maintainability of systems.[13] These observations lead to our third recommendation: *In seeking to reap the benefits of commercial software development practices, DoD must take care not to replicate the security practices of that industry.*

We categorized systems as safety-critical if they were highly autonomous or if they could cause harm to their users.[14] We found that roughly 65 percent of DoD major acquisition programs are safety critical. In developing software that provides command and control of safety-critical systems, software professionals must specify and implement methods, processes, tools, and models (the elements of a safety culture) to identify, mitigate, and/or remove hazards from the acquired software and the systems they support. These methods and processes for developing safety-critical software are well established in the medical device, nuclear reactor, and aeronautics industries and, to a lesser degree, within DoD.[15] However, applying the principles of safety engineering within an iterative software development lifecycle—as is currently being advocated across DoD—is not straightforward.[16] Although it is possible, doing so requires impeccable planning, change management, and documentation to address regulatory compliance.[17] It is likely that even the most competent of DoD safety-critical software developers will need additional training if they are to be successful within an iterative development lifecycle.

DoD is increasing the automation of many of its weapon systems, which necessarily introduces the need for additional safety standards. Our review of aircraft industry and DoD's standards for developing safety-critical software did not find adequate coverage of methods and processes needed to ensure the high levels of operational

---

[13] Deploying microservices within security-hardened containers is a best practice that is currently being pursued within DoD. An example of this practice is discussed by Nicholas Chaillan, U.S. Air Force chief software officer, in Tom Krazit, "How the U.S. Air Force Deployed Kubernetes and Istio on an F-16 in 45 Days," *New Stack*, December 24, 2019.

[14] Although the argument could be made that a failure to effectively engage an adversary could cause harm to the user of any warfighting system, we classified battle management and logistics systems as "not safety critical." For sensors, radars (with their ability to emit high-powered and damaging radio frequencies) were classified as safety critical, while more-passive sensors (such as infrared) were not. Most communication systems were also classified as not safety critical. Absence of communications can lead to loss of life on the battlefield, but that loss of life is not caused by use of the communication device.

[15] For the aviation industry, Leanna Rierson has written an excellent guide (Leanna Rierson, *Developing Safety-Critical Software: A Practical Guide for Aviation Software and DO-178C Compliance*, Boca Raton, Fla.: CRC Press, 2013. DoD's standard for the development of safety-critical software, MIL-STD-882E (last updated in May 2012), is augmented by implementation guidance in DoD, *Software System Safety: Implementation Process and Tasks Supporting MIL-STD-882E*, Washington, D.C., Revision A, October 17, 2017.

[16] See Appendix B in Robson et al., 2020, for an overview of the agile development lifecycle and some of the challenges of using agile practices within DoD.

[17] For a tutorial on applying iterative development lifecycles to safety-critical systems, see Nancy Van Schooenderwort and Brian Shoemaker, *Agile Methods for Safety-Critical Systems: A Primer Using Medical Device Examples*, Lean-Agile Partners and ShoeBar Associates, 2018.

trust critical to autonomous systems.[18] Current efforts to develop DoD standards for autonomous weapon systems focus on the validation and verification of those systems; although these efforts address security concerns, they are noticeably silent on safety issues.[19] This leads to our fourth recommendation: *DoD should review current safety standards and guidance to ensure they include best practices for achieving high levels of operational trust in highly automated and/or autonomous DoD software-intensive systems.*[20]

---

[18]   The standards reviewed were DO-178C, MIL-STD-882E, and DoD Directive 3000.09.

[19]   The DoD standard governing autonomous weapons is DoD Directive 3000.09, *Autonomy in Weapons Systems*, Washington, D.C.: U.S. Department of Defense, Incorporating Change 1, May 8, 2017. It includes a reference to DoD's development standards for cybersecurity but not to DoD's development standards for safety.

[20]   Recommended readings regarding Validation and Verification methods and processes to assess operational trust in autonomous systems include Joseph B. Lyons, Matthew A. Clark, Alan R. Wagner, and Matthew J. Schuelke, "Certifiable Trust in Autonomous Systems: Making the Intractable Tangible," *AI Magazine*, Vol. 38, No. 3, Fall 2017; Raj Gautam Dutta, Xiaolong Guo, and Yier Jin, "Quantifying Trust in Autonomous Systems Under Uncertainties," *Proceedings*, 2016 29th IEEE International System-on-Chip Conference (SOCC), Seattle, Wash., September 6–9, 2016; Jan Hodicky, "Autonomous Systems Operationalization Gaps Overcome by Modelling and Simulation," in Jan Hodicky, ed., *Modelling and Simulation for Autonomous Systems*, Third International Workshop, MESAS 2016, Rome, Italy, June 15–16, 2016; and Saeid Nahavandi, "Trusted Autonomy Between Humans and Robots: Toward Human-on-the-Loop in Robotics and Autonomous Systems," *IEEE Systems, Man, and Cybernetics Magazine*, Vol. 3, No. 1, January 17, 2017.

CHAPTER FOUR

Challenges Faced by Software Development Organizations Within DoD

To explore the challenges facing software development organizations within DoD, we interviewed program managers (and often their chief engineers or software leads) to understand how software talent is employed, barriers they are encountering, and any best practices they might want to pass on to their counterparts in other organizations. Originally, we hoped to interview a representative sample, as determined by the distribution of program system types across the FY 2018 major programs (described in Appendix A). Unfortunately, we were unable to secure a task memo from the Office of the Secretary of Defense (OSD) or from the service acquisition offices to facilitate our data collection efforts. Instead, we solicited participation using the personal contacts of the sponsor and of our research team to perform a "snowball sample."[1] Our interviews were heavily represented by Air Force, Joint, and enterprise system programs (Tables 4.1 and 4.2). Therefore, this analysis may not be generalizable to the Navy and embedded systems that dominate DoD procurements (see our analysis of major programs docu-

Table 4.1
Service Branch of Interviewed Programs Versus FY 2018 Major Program Software Value

Service Branch	Percentage of Interviews	Percentage of FY 2018 Major Program Software Value
Army program	13	16
Navy program	13	65
Air Force program	38	18
Joint program	25	2

NOTE: The highlighted row emphasizes the nonrepresentativeness of our interviews. Although we believe the observations in this report offer insight, we cannot claim extensibility across DoD.

---

[1] A snowball sample starts when researchers select a small number of willing participants and then use those participants to recruit others. The result is a nonrepresentative sample and is likely biased is unknown ways. This sampling method makes it impossible to generalize results beyond the sample population. However, snowball sampling does allow us to describe characteristics and details about the sample.

Table 4.2
System Type of Interviewed Programs Versus FY 2018 Major Program Software Value

System Type	Percentage of Interviews	Percentage of FY 2018 Major Program Software Value
Embedded system	38	95
Enterprise system	63	5

NOTE: The highlighted row emphasizes the nonrepresentativeness of our interviews. Although we believe the observations in this report offer insight, we cannot claim extensibility across DoD.

mented in Chapter Three). Details regarding our data collection methodology for this task, including the interview protocol, are provided in Appendix A.

Managing the Change to Modern Software Practices

Adoption of Agile Practices Is On the Rise

All program offices we interviewed noted that they already have agile or agile-like practices in place or are developing strategies to transition to agile practices in the foreseeable future.[2] In almost all cases, however, the agile development tempo chosen for release of working DoD software is measured in months rather than the days or weeks used in commercial industry. As noted in our prior report (Robson et al., 2020), a key competency for DoD software professionals is the synchronization of software incremental releases with the underlying, and possibly evolving, hardware. As we might expect, programs achieving shorter software delivery tempo are generally those with more-stable underlying hardware.

Some offices noted, however, that DoD's current workforce may not all have the necessary mindset to thrive on the cutting edge of software development practice. Agile practice in commercial industry emphasizes the gains that come from experimentation and learning from failure; it is a bit cliché to say government bureaucracies do not. One seasoned manager of a DoD program that had fully transitioned to agile practices estimated that they had experienced an almost total turnover of staff in the

---

[2]   See Appendix B of Robson et al., 2020, for a more complete description of the trends in modern software development that are colloquially known as "agile." Agile is a perhaps overused term in software circles. As used here, it includes adoption of the principles in "The Agile Manifesto," which are most concretely realized by the adoption of an iterative development lifecycle and various practices, such as test first, pair programming, continuous integration, continuous development, and continuous deployment. Specific implementations of these practices go by various names, such as Extreme Programming, SCRUM, Kanban, and DevOps. The Agile Manifesto is a short document written and signed in 2001 by 17 leaders in the software industry, who outlined four values and 12 principles that are essential for improving software development practice (Kent Beck, Mike Beedle, Arie van Bennekum, Alistair Cockburn, Ward Cunningham, Martin Fowler, James Grenning, Jim Highsmith, Andrew Hunt, Ron Jeffries, Jon Kern, Brian Marick, Robert C. Martin, Steve Mellor, Ken Schwaber, Jeff Sutherland, and Dave Thomas, "Manifesto for Agile Software Development," webpage, 2001).

process. They stated that the staff turnover came in two waves: an initial wave as personnel experienced discomfort in being asked to adopt modern software development practices and a later wave when willing participants found they did not have the skill set or aptitude to effectively implement those practices.

Programs Find It Difficult to Make the Voice of the Customer Central to Software Development

A guiding principle of modern software development practice is that the voice of the customer must be central to the development effort. Obtaining customer input should be accomplished by engaging end users early and often in the process, not just to explain and validate the desired features of the software but to continually participate with the team in setting priorities for which features to develop next and assessing whether delivered features are actually useful and will meet the operational need. Ideally, DoD operations personnel would periodically rotate through acquisitions, bringing recent experience to this task.

However, program managers report continued difficulties in finding individuals who are qualified to fill these voice of the customer roles. They report significant cultural barriers, suggesting that improving the competency of warfighters in software development practice is unlikely to have an impact without accompanying cultural change. Even when a program is in direct support of an operational command, program managers report that there is reluctance to ask personnel to take time off from operations to support software development. More often than not, the chief engineers or lead software engineers admit to attempting to fill this void themselves. None believe they are doing it as well as it should be done. DoD's independent operational test and evaluation (IOT&E) personnel are well suited to perform these roles. In fact, it is their charter to assess whether delivered products meet operational need. From our conversations, there appear to be at least two barriers to IOT&E taking a more active voice of the customer role: (1) IOT&E's perception that becoming more tightly integrated into the development of the software would compromise their independence and (2) program management's perception that IOT&E's independence will drive additional change into the development that is out of the program manager's control.

Chief Engineers Rarely Have Experience in Software Development

Chief engineers do not appear to be present in all program offices that are concerned with software acquisition. Where they do exist, the chief engineer often lacks computer science or software expertise. This observation mirrors our statistical analysis of the snapshot of the workforce data in Chapter Two, which showed that those who manage software often do not have an educational background in software. The chief engineers we interviewed stated that they focus primarily on systems engineering efforts. Interviewees described the chief engineer role as a strategic coordinator: someone who should organize, train, and equip staff for their jobs while providing a unifying vision.

Their role is made even more difficult by the challenge of having to coordinate across several different acquisition programs and staff with diverse skill sets.

Although the emphasis of the chief engineers on system engineering is undoubtedly necessary, performing the chief engineer role on a software-intensive program without a background in software development is challenging. Although no single person can have a deep background in all technical areas of a program, it is rare for a chief engineer to have no background in hardware development. The fact that it is common for them to have no background in software development, despite the hundreds of millions of dollars at risk if that software is not developed appropriately, should give us pause. Given the rapid evolution of software skills and tooling, it may be impossible to perform the train and equip task without being current, or having access to trusted expertise that is current, in software development methods. Staying current in the field is a significant challenge, as we will discuss later. As for the role of strategic coordinator, software developers within DoD make thousands of decisions every day that will affect the way that software operates (or fails to operate) in the field, the way it is delivered, and the way it is sustained. Understanding which of those decisions will affect the overall program is a critical skill set for the program's chief engineer.

For complex systems, it is not uncommon in the commercial workforce to employ an Office of the Chief Engineer, comprised of a small number of senior engineers (typically three to ten) who together have deep expertise in the different technologies of the system (for example, the chief engineer for a missile warning system needs expertise in guidance, navigation, and control, optical and radio frequency sensor networks, and both embedded and enterprise software—a rare combination to find in any single person).[3] Interestingly, we did not find a similar concept of shared chief engineer accountabilities in any of the programs we interviewed. Although this may be due to our small sample size, we noted that many of the programs we interviewed had a lead software engineer who appeared to work closely with the chief engineer. It may be that those programs were operating in a shared accountability manner as best they could, given DoD's strong culture of hierarchical management structure.

Just-in-Time Training Is Critical but Not Always Available to Personnel

As we noted in our prior report on competencies for the DoD software acquisition workforce, best practices in software development change rapidly. Personnel are not always familiar with modern software development practices (such as DevOps), necessitating participation in continual trainings to keep skills and knowledge as current as possible. Furthermore, more than one of our interviewees noted that training that is not immediately put to use is often forgotten. Therefore, periodic training is needed to

---

[3]   In the automotive industry, the group of Level 1 Chief Engineers are sometimes designated the "vehicle program steering team" (Vivek D. Bhise, *Automotive Product Development: A Systems Engineering Implementation*, Boca Raton, Fla.: CRC Press, 2017).

maintain familiarity, but just-in-time training is critical during the development and deployment of systems.

Software personnel, however, may not always be aware of what training is available, appropriate, or adequate. Some interviewees mentioned that their program offices brought in coaches or other instructors to speak on specific topics, such as development of a pipeline for continuous integration and/or deployment of software, but that practice did not appear to be standardized. As a result, some software personnel reported that they often seek out their own trainings online via platforms such as YouTube.

Leaner program offices may face an additional barrier to providing training opportunities to their staff. Interviewees said that a lack of funding may hamper program offices' abilities to pay for training courses.[4] Likewise, offices with staffing shortages may not be able to part with employees in critical positions for lengthy periods of time. Discussions about how to ensure strategic investment in staff development do not appear to take place regularly.

Interviewees also cited other organizational and cultural barriers to training. For instance, obtaining outside training (often supplied by tool developers or commercial partners for a fee) may require support from more-senior officials. Those officials, one interviewee said, may have a bias against additional software-specific training because it is seen as the "new fad" in DoD. Further, although many interviewees said they knew of software courses available through the Defense Acquisition University (DAU), they also pointed out that technical training is the responsibility of the military services education organizations. More than one interviewee suggested that a remedy to these issues may be found through more collaboration between DAU, the service training organizations, and commercial partners to access training available in the private sector. Unfortunately, this may not be as straightforward as some imagine: Not all private sector training will take into consideration the security- and safety-critical practices needed for DoD software development.

In a May 2019 report, the Defense Innovation Board (DIB) noted many of these same training issues regarding DoD's software workforce. In its report, the board noted that although the current training model for acquisition professionals has helped to build the world's best military, it is not serving well for software. In particular, the board said,

> New methodologies and approaches introduce unknown risks, and acquisition professionals are not often incentivized to make use of the authorities available to implement modern software methods. At the same time, senior leaders in DoD need to be more knowledgeable about modern software development practices

---

[4]   It is our observation that the issue our interviewees reported was perhaps not a lack of funding but rather a lack of funding designated for discretionary training. Required training often consumes an office's entire training budget.

so they can recognize, encourage, and champion efforts to implement modern approaches to software program management.[5]

The DIB recommended the expansion of specialized training to provide more insight into software development practices and authorities to enable rapid software acquisition. The DIB also recommended modification of U.S. Code and the establishment of a software acquisition workforce fund, similar to the Defense Acquisition Workforce Development Fund, to be used for hiring and training software acquisition experts.[6] We note, however, that allocating funds for such an initiative is difficult if we cannot answer such basic questions as who the software acquisition personnel are and what their expertise and current pay scale are.

Managing Software Development

DoD Software Acquisitions Are Highly Dependent on a Large Contractor Workforce, Overseen by a Small Cadre of DoD Personnel

Our first observation regarding the organizational challenges of developing software within DoD is that the workforce in the direct employ of DoD is vastly outnumbered by the contractors who work for them. Ratios of 1:40 or higher are the norm, not the exception.[7] Despite a recent push to create and use "soldier coders," most software within DoD is still developed by contractors.[8] This imbalance suggests that software project, program, and contract management are critical skill sets required of DoD's software workforce.

To gain perspective as to whether DoD is appropriately training personnel in these management skills, we revisit our prior work developing a competency model for this workforce.[9] In support of that effort, we hypothesized possible career paths within the workforce, two of which were program manager and software project manager while the others focused on more-technical roles. Although our prior work mapped desired competencies to those roles, for this report, we trace not just from roles to com-

---

[5]   Defense Innovation Board, 2019, p. S36.

[6]   Defense Innovation Board, 2019, p. S36.

[7]   This observation is based on our interviews with the program offices and was further supported by conversations with researchers conducting two similar DoD software–related studies and analysis efforts.

[8]   In 2016, Maj. Gen Bruce Crawford estimated that the Army relies on contractors for about 85 percent of the effort required to acquire and sustain the software in Army systems (Jared Serbu, "Army Asks How Many Civilians, Contractors It Needs for Software Development of Weapons Systems," *Federal News Network*, October 10, 2016).

[9]   Robson et al., 2020.

petencies but also to the DAU classes that support those competencies.[10] What we find is that DAU classes are well-suited to train project management skill sets but fall short in covering the broader skills needed to understand the technical aspects of software (Table 4.3). DAU courses are particularly unsuited to developing skill sets needed by enterprise and software architects and by software technologists.

Understaffed Program Offices and a Lack of Desired Skill Sets in Workforce Contribute to a Dependency on Contractors

Our second observation is that this dependency on contractors is at least partially driven by an understaffed and underskilled software workforce. Several of our interviewees reported that DoD software personnel do not fully understand the systems they are developing or the effort needed to code and build the systems, ensure cybersecurity is sufficiently assessed, field the system, and train end users. Deficiencies in program knowledge and critical skill sets can lead to a significant disconnect between DoD's software personnel and the programs they are developing or overseeing. For uniformed personnel, the lack of program knowledge can be partially attributed to turnover: The average duration for uniformed staff on any given program is less than three years.[11] Nonuniformed personnel within DoD spend a longer time on a given program, allowing them to build greater program expertise. However, a 2019 GAO report found that the DoD programs they examined had failed to upgrade software

Table 4.3
Percentage of Competency Requirements Fully Met by DAU Training Courses

Software Role	Percentage of Primary Skill Sets Covered by DAU ISA Courses
Software project managers	100
Software integration managers	92
Program managers and system engineers	89
Enterprise and software architects	78
Software technologists	56

SOURCE: RAND analysis based on data collected and reported in Robson et al., 2020.
NOTE: ISA = information systems acquisition.

---

[10]   The mapping of roles to competencies is shown in Appendix D and of competencies to DAU courses in Appendix H of Robson et al., 2020. Table 4.3 was created by mapping the primary competencies needed for each role to training courses that fully covered the competency.

[11]   High turnover of personnel in complex software-intensive acquisition programs is not a new issue. A 2015 review of the troubled Global Positioning System Next-Generation Operational Control (OCX) system cited high personnel turnover as one of five contributing causes of that program's difficulties. Unfortunately, this is still the case in 2020. For more information on the OCX program, see Thomas Light, Robert S. Leonard, Meagan L.

development tools and methods over the life of the program, meaning that a longer tenure may also allow software skill sets to atrophy.[12]

Many offices stated that they relied heavily on their contractors to bridge this disconnect. The level of dependency varied, but more than one program office noted a total dependency on qualified contractors, stating that personnel directly employed by DoD lacked the technical background to be software engineers or do the necessary coding. Other program offices reported having a highly qualified and highly trained workforce but felt they were overworked and understaffed.

Reliance on Contractors Is Hampered by Difficulties in Administering Contracting Vehicles

During our interviews, we asked programs to share lessons learned. Although the answers we received were not specifically related to competencies, we collected several stories about how DoD's current contracting processes hinder programs seeking to access contractor skill sets. To support modern software development methods, agility in contracting to adjust scope and to deal with a wider diversity of nontraditional DoD contractors is needed. One program office shared an insightful comment that traditional DoD contracts are not written to support software acquisition because contracts are not allowed to grow. Shifting to different contract vehicles, such as indefinite delivery/indefinite quantity, was identified by interviewees as a practice that may help software acquisition program offices construct contracts that can adjust to changing needs.

Other interviewees noted that contracts can take an inordinate amount of time to get approved. Some singled out the Fed Biz Ops system, now known as Contract Opportunities on the beta System for Award Management website, for its slowness. As a concrete example of the slowness of DoD contracting, another program office noted it took about a year to have a service-level agreement—a typical method of contracting for cloud services (i.e., infrastructure, platform, or software as a service)—written, signed, and approved. One interviewee also noted that there was no good way for DoD to partner with smaller contracting firms, which do not typically have the infrastruc-

---

Smith, Akilah Wallace, and Mark V. Arena, *Benchmarking Schedules for Major Defense Acquisition Programs*, Santa Monica, Calif.: RAND Corporation, RR-2144-AF, 2018.

Turnover of program manager tenure is cited as being every two to three years by the service chiefs in a 2015 GAO report. Although the service chiefs stated that this makes it difficult to hold managers accountable when program issues emerge, it seems to us that it is equally likely that these short tenures make it difficult to achieve an adequate understanding of the technical and managerial complexity of software-intensive systems (GAO, *Defense Acquisition Process; Military Service Chiefs' Concerns Reflect Need to Better Define Requirements Before Programs Start*, Washington, D.C., GAO-15-469, June 11, 2015).

[12] GAO, 2019.

ture to meet many DoD contracting requirements. All of these comments are consistent with the 2019 findings of the DIB.[13]

Acquiring Software Talent

Difficulties with Hiring and Retaining Software Personnel

Difficulties with hiring government employees contribute to contractor dependency and a lack of knowledge in the DoD software workforce. Several interviewees mentioned a governmentwide shortage of personnel with software development experience. This shortage will likely only get worse over the next decade because nationwide demand for software developers is projected to grow 21 percent between 2018 and 2028, much faster than the average across other occupations.[14] Congress has taken note of how this skill shortage may affect DoD, urging defense leaders to create a pipeline from top universities into the Pentagon. However, that pipeline will need strong incentives to attract the best recruits, particularly because the private sector can offer more attractive salaries for experienced and entry-level software developers. One news article notes that the average University of Illinois computer engineering graduate in 2018 earned a starting salary of $99,741, nearly $600 more than a GS-13 Step 1 employee in the Washington, D.C., area.[15] Although this initial difference in salary is not large, our analysis comparing the DoD software workforce to that of the industry workforce, documented in Chapter Two, suggests that salary disparities grow significantly over the course of a career.

Cultural factors also affect hiring and retention. One interviewee said that few people with upper-level software engineering (or other computer science) degrees want to do DoD acquisition work. Expectations of those upper-level personnel may also be unrealistic. Interviewees noted a dearth of qualified individuals in the workforce who have experience in acquisition, industry, and the government. There was an assumption that effective software personnel need knowledge of the entire acquisition lifecycle; experience in the private and public sector; an ability to develop, code, test, and ensure the security of software programs; and an ability to lead or oversee contractors who partner with their program office. This reflects similar findings of a 2019 survey of government CIOs and IT leaders, who sought not just technical skills in new recruits but a combination of interpersonal skills and interdisciplinary backgrounds. Those skills included communication, cognitive flexibility, people management, and complex problem-solving. Other skills requested included knowledge of data science,

---

[13] Defense Innovation Board, 2019, p. 12.

[14] U.S. Bureau of Labor Statistics, "Occupational Outlook Handbook: Software Developers," webpage, accessed June 10, 2020.

[15] Patrick Kelley, "The Pentagon Is Facing a Serious Workforce Problem," *Government Technology*, July 10, 2019.

business and economics, marketing, and engineering.[16] We note that this is a lot to ask of any one individual.

Some program offices also indicated difficulties with getting members of their workforce coded appropriately. Two offices outlined struggles getting staff coded to reflect their software development skills or responsibilities: One office mentioned difficulties getting staff coded as software engineers, and another highlighted difficulty getting staff properly coded as computer scientists because of a common view that "acquisition programs do not need software developers." This view is one the DIB challenged in a 2019 report, saying that DoD

> cannot compete and dominate in defense software without a technical and design workforce within the Department that can both build software natively and effectively manage vendors to do the same, using the proven principles and practices described [in another part of the report]. Some of the Department's human capital practices work against this critical goal.[17]

The report also highlights best practices from private sector organizations, including understanding the skill sets of the currently employed software professionals, understanding the skills the workforce must perform to be successful, and understanding the gap between the two.[18]

A more thorough examination of current human capital practices could give DoD a better sense of what could be done to improve recruiting and retention of younger software engineers, including the benefits and opportunities they may seek out (career tracks, flexibility, training and opportunities for development, etc.). Although we found anecdotal evidence of difficulties in hiring software engineering talent, DoD will face challenges in conducting more-quantitative analyses. There are no official positions in software engineering or software development, and our search of government databases found no other indicators to track which offices or programs are having difficulty hiring software talent. One program manager, discussing DoD's talent acquisition ecosystem, estimated that it may take upwards of four years to build the right policies and processes and to implement the needed updates to modernize that system.

Software Acquisition Requires a Broad Variety of Skill Sets

Our third observation regarding the challenges in managing software development is that personnel with a diverse skill set are highly valued. As documented earlier, program managers expressed a need for software staff with knowledge of acquisition processes, the software environment, coding, trends in industry and private sector, and

---

[16] Workscoop and Fedscoop, "Reskilling the Federal IT Workforce," undated, p. 7; Phil Goldstein, "Which IT Skills Are Most in Demand in Federal IT?" *FedTech*, January 6, 2020.

[17] Defense Innovation Board, 2019, p. 12.

[18] Defense Innovation Board, 2019, p. 12.

knowledge of the system or systems being developed and acquired. Staff should also be flexible and adaptable, able to overcome complications that may arise from a mix of commercial-off-the-shelf hardware and software, as well as custom software. These desired ideal skill sets take time to develop and require a variety of experiences. Further, program offices stated software personnel might lack a technical background. As discussed in Chapter Three, DoD software is dominated by embedded systems that are tightly coupled to system hardware and thus require software engineers who can bridge the technical gap to that greater system.

Lack of a communitywide list of skills and competencies that can be referred to or edited over time might also contribute to the perceived lack of skill sets. In fact, individual positions in the software community that may appear to have similar responsibilities and tasks are very likely to require different competencies and should be filled by individuals with different backgrounds.[19] For example, a web applications developer needs or has very different skill sets than a safety critical embedded systems developer, although both are software developers. Being able to distinguish between those types of developers for recruiting and training requires a canonical list of software competencies from which to select the specific skill sets needed by a program.

---

[19] The differences between designing a web app versus embedded software is analogous to the difference between mechanical and electrical hardware systems. As the software industry matures, specializations such as these will become more distinct. Software engineering educational programs already have begun to specialize in particular genres of software development. One online resource for college applicants partitions software career paths as applications development, systems development, web development, and embedded systems development, but other categorizations are also common. See ComputerScience.org, "Exploring Software Engineering: A Comprehensive Guide to Careers and Top Employers," webpage, accessed June 2020.

CHAPTER FIVE

DoD Workforce Data Call Lessons Learned

Characterizing the DoD software workforce requires identifying personnel who perform software activities in DoD. However, we found no specific identifiers or combination of data attributes within the existing DoD manpower and personnel systems that can be used to reliably identify who does or does not perform software activities. This lack of identifying information was our primary challenge in characterizing the workforce. Thus, collection of new data was required to identify personnel who perform software activities. In this chapter, we describe the data collection objectives and tasks carried out by the research team. We also provide some valuable recommendations and lessons learned for future data collection activities of a similar nature.

Data Call Objectives: Constrained Options for DoD Software Workforce Data Collection

The initial desired end state for data collection was to execute a full census of DoD's IT, engineering, program management, system engineering, test and evaluation, and selected logistics personnel to identify who performs software activities and then merge the collected data with existing DoD personnel databases to provide a complete picture of that workforce. Our original intent was to field a simple one-question survey for up to 50,000 uniformed and nonuniformed DoD personnel asking whether they performed activities related to the acquisition of software and then add that information to existing personnel data records. However, the options for achieving this outcome became increasingly limited during the planning and implementation phases of the data call. As discussed in the following sections, the reasons included a reorganization of the sponsoring agency, changes among key government stakeholders, and a lack of support for the data call from some DoD service organizations. Instead of a full census, we collected data from only a small sample of software personnel in organizations with known concentrations of software personnel.

Data in Existing Systems Cannot Identify Software Personnel

The research team began planning for the data call by seeking and examining existing sources of personnel data to determine whether they contained all or some of the target data and to determine the processes and feasibility of modifying existing systems to receive the data. The following systems were examined:

1. The DCPDS
2. The Acquisition Career Management Systems for each service:
   a. Navy Electronic Defense Acquisition Career Management
   b. Air Force Acquisition Career Management System
   c. Army Career Acquisition Management Portal/Career Acquisition Personnel and Position Management Information System
   d. Fourth Estate Defense Acquisition Talent Management System
3. The Defense Manpower Data Center (DMDC), Active Duty and Civilian Files
4. Data Mart.

Many of the existing databases share data with each other and have common data fields that we hoped to exploit for our research. Although much of the organizational, demographic, and career management data were available, our examination of the existing systems confirmed there were no software-specific codes or any combination of attributes distinct to software activities that could be used to accurately identify individuals who perform software-related activities in DoD. Although position titles in DCPDS and in the other systems that contained data on nonuniformed positions were descriptive of the personnel's occupational series, there is no existing software-related series.[1] Similarly, there is no acquisition career field specific to software development.[2] A review of special skill identifiers available in various databases also found no useful means of identifying personnel involved with the software-related activities. Although there are some distinct skill identifiers in the data dictionaries for some of the services that led us to hope we would find meaningful data, we found that those fields were not

---

[1]   Occupational series designators are used by the U.S. Office of Personnel Management and are common across the U.S. government. They are used to define the knowledge required for personnel in those positions. Although there is a computer engineering series, a computer science series, and an IT series, none require detailed knowledge of software acquisition, development, and sustainment practices. Although creating a new occupational series is a highly politically charged undertaking, we believe updating these designators to include software engineering may be necessary if the U.S. government is to harness the full potential of modern software acquisition practices, given the prevalence of software across the U.S. government.

[2]   Acquisition career field designators are used by DoD to establish training and certification requirements for acquisition professionals. Prior to our research, we hypothesized that software personnel would be concentrated in the existing career fields of IT, engineering, program management, test and evaluation, and logistics, but we had no evidence to support that hypothesis. Using our findings in this study, we can now say the bulk of software acquisition personnel are in the engineering career field. We believe there would be benefit in creating a subspecialty within that career field for software engineering to emphasize the distinct competencies needed by software acquisition professionals. For an overview of those competencies, see Robson et al., 2020.

populated or were sparsely populated after we gained access to the data. To understand whether sparse data was relevant, we pulled records from geographic locations where we knew the service had a large contingent of software personnel. A lack of data for these centers indicated that these fields were not a reliable indicator.

A particular frustration was the lack of educational background in DMDC data. Although the DMDC data dictionary includes a field for educational discipline code, when we were granted access and pulled records for software data centers, we found this field largely blank, providing us no insight into the educational background of the personnel working in those centers. Even if those fields had been populated, we are not optimistic that it would have provided the insight we need for our work because personnel are not always employed in their field of study and because a review of the data dictionary reveals that although you can be educated as a "card dealer" and in everything from art therapy to zoology, educational discipline codes do not include the breadth of disciplines applicable to software acquisition. Computer science and computer engineering are included, but software engineering and cybersecurity are not.[3]

Options Considered for a DoD-Wide Software Workforce Data Call

After examining options within existing databases for identifying the software workforce, the research team considered several options to execute a DoD-wide data call to obtain the information we needed. Table 5.1 lists those options along with the advantages and considerations associated with each.

After analyzing the existing systems and conducting initial planning sessions with the research sponsor's leadership and other stakeholders, the research team decided to pursue the modification of the acquisition career management systems to add new software workforce identifiers, followed by the coordination of a data call task memo delivered from the Office of the Under Secretary of Defense for Research and Engineering (OUSD[R&E]) to DoD component heads, acquisition career managers, and functional career leaders requesting the update of the modified systems to identify software personnel.

Lack of Stakeholder Support for a DoD-Wide Data Call

Stakeholder outreach for the data call included the establishment of a data call working group that introduced the initiative and data call requirements to stakeholder organizations. The purpose in forming the working group was to assess support and iden-

---

[3]   Although there are debates regarding how and why software engineering is different from computer science or computer engineering (and why cybersecurity is different still), there is broad consensus that they are different. Computer science is the scientific study of how data is organized and processed, the design and security of computer networks, and the development of websites and applications. Computer engineering is the application of engineering principles to computer design and development. Software engineering is the application of engineering principles to software design and development. Cybersecurity is the application of computer science principles to the security of software, computers, and networks.

Table 5.1
Data Collection Options for the Software Workforce

Methodology	Advantages	Considerations[a]

Spreadsheet data collection—tasked by DoD components, functional leads, or acquisition career managers	• Requires minimal IT Infrastructure
	• Familiar data entry function
	• Responsibility for task completion mostly on respondent organization
	• Can send spreadsheet to anyone in DoD (or any subset)
		• Higher risks of:
		  • Invalid data (nonstandard entries)
		  • Human error
		  • Personally identifiable information mishandling
		  • Delays and timeline revisions
		  • "Lost" or overwritten files
		• Difficult to track completion and response rate
		• Possible duplicate responses
		• Data utility beyond DCAT and RAND study diminishes

Online survey sent to organizational distribution lists	• Data validity is high • Easily tracked response rates • Ability to follow up with missing responses	• Requires email testing for each respondent organization • Timeline pressure because of possible OPM review and approval of survey instrument • Self-reported data • May require SORN and PIA • Data utility beyond DCAT and RAND study diminishes
Modify DoD personnel systems—DoD components task individuals to update their own records	• Familiar data entry • Use of existing systems and process • High data validity • Responsibility for task completion on respondent organization • Can collect data for all DoD personnel • Database can continue to track workforce after data call complete	• Difficult to track completion and response rates (false negatives) • Separate coordination required for each system to update specialty codes (changes required to at least 5 databases) – differing change & control processes for each system • Scheduling of system enhancements & data transfers outside Government Sponsor's direct control • Competing system modification priorities with other DoD entities
Modify acquisition career management systems—acquisition career managers task components to identify software personnel	• Minimal IT development • Familiar data entry functions • Use of existing systems and processes • Very high data validity • Responsibility for task completion on respondent organization • All ACM systems use similar technology and change and control processes	• Difficult to track completion and response rates (false negatives) • Separate coordination required for each system owner to update specialty codes (four ACM systems and, possibly, Data Mart modifications) • Scheduling of system enhancements and data transfers outside government sponsor's direct control • Excludes personnel not in acquisition-coded positions

SOURCE: Country Intelligence Group analysis.
NOTE: ACM = acquisition career management; DCAT = Data Catalog Vocabulary; OPM = Office of Personnel Management; PIA = privacy impact assessment; SORN = system of records notice.

[a] Note that considerations are not necessarily disadvantages: They simply are factors that need to be taken into account when considering the limitations of the data collection method and how that might affect the validity of subsequent research that uses the data.

tify additional opportunities and challenges for the data call. The working group was chaired by the OUSD(R&E). Working group invitees included all four defense acquisition career management (DACM) offices, DoD's human capital initiatives office, the Defense Civilian Personnel Advisory Service, and representatives from the DAU, Defense Logistics Agency, DIB, OSD Chief Information Office, Office of the Under Secretary of Defense for Acquisition and Sustainment (OUSD[A&S]), and Defense Information Systems Agency. Representatives from all invited organizations were able to attend one or more of the working group meetings except for the Navy DACM. The Navy provided input and feedback to the data call team outside of the working group discussions. The Navy's feedback was generally unsupportive of our efforts. The working group disbanded after three meetings with the conclusion that although a data call is technically possible, there was insufficient support across the DoD for a full census data call. The lack of support could be attributed to several factors, including

- fatigue with data calls or surveys across DoD
- inadequate resources within organizations to facilitate data collections
- lack of a perceived benefit to service acquisition executives from data calls and, for some, a belief that software should not be singled out separately from other types of acquisition
- expectations of key stakeholders that low response rates would invalidate the effort
- desires for a data collection tool other than a spreadsheet, which was felt to be error prone and inefficient, or for direct entry into existing systems, which (although efficient) would not allow for independent vetting of data prior to entering it into authoritative systems[4]
- the need for a task memo directing the data call from a level at or higher than Under Secretary of Defense[5]
- the pursuit of competing priorities by software supervisors, which results in them having too little time to coordinate data collections in their centers.

Fallout from Government Sponsor Agency Reorganization
At the start of the project, our sponsor was aligned under the Office of the Under Secretary of Defense for Acquisition, Technology and Logistics (OUSD[AT&L]). Shortly thereafter, OUSD(AT&L) was reorganized into two new entities: OUSD(A&S) and OUSD(R&E). Although our sponsor now fell under OUSD(R&E) and no longer had accountability for the acquisition workforce, that office is affected (positively or negatively) by the capabilities of that workforce and continued to support our research, as

---

[4]   We did design such a tool in collaboration with OUSD(R&E)'s Mission Engineering office. See the next section regarding organizational issues that complicated efforts to implement this tool in a timely fashion.

[5]   Other DoD-wide data calls for functional matrixed workforces (e.g., the DoD Security Cooperation Workforce) have been successful in achieving high response rates from task memos coordinated at the DoD-component-head level and then signed out at the Under Secretary of Defense level.

did officials from OUSD(A&S).[6] Despite this support, the residual effects of the reorganization resulted in vacated leadership positions within the sponsor's office and the realignment of support services between the two new entities, each of which affected our work in different ways, as follows:

- Vacated leadership positions left the data call effort without a government representative to introduce the initiative to other stakeholder organizations and software centers. This also affected the ability to coordinate a data call memo at the appropriate level to enable a DoD-wide data call.
- After the reorganization, the Mission Engineering office that previously provided technical development services to OUSD(R&E) was now aligned under OUSD(A&S). Although we were near completion of our collaboration with the Mission Engineering office to develop a new, yet simple, automated data collection tool, as requested by the data call working group, Mission Engineering was no longer able to task the technical service support office they had traditionally worked with. Although technical support services eventually became available to OUSD(R&E) through another office, this handover created a backlog of other unrelated requirements, which the data call was competing against. Ultimately, tool development proceeded too slowly to be considered a viable option for timely data collection.

Eventually, both the vacated leadership and lack of technical services were addressed, but not before DoD announced an indefinite pause on all functional career area–related activity while an assessment was conducted on the functional areas. Therefore, there was little the functional lead cosponsors of the study (IT and Engineering) could do to facilitate and promote a DoD-wide data call. Facilitation issues were later resolved by the government sponsor with the addition of a government representative to the project team, detached from the Army's Combat Capabilities Development Command (CCDC)/Armaments Center.

---

[6]   At no time did the research team feel that we did not have the support of both OUSD(R&E) and OUSD(A&S) leadership. As the DIB has pointed out, software development stretches across all phases of development—research, acquisition, engineering, and sustainment—and modernizing practice is one of the larger challenges DoD faces. However, the split created bureaucratic hurdles that could not be overcome without the expenditure of significant political capital. As only one of many competing issues of concern within DoD, we found ourselves unable to marshal support at the level needed to make the data call a reality.

Leveraging Working Group Products to Build a Coalition of the Willing

Although the data call working group did not produce a coordinated and signed data call task memo, there were several productive outcomes from the working group. These include the following:

- The development of a software activity model that could be used to identify software personnel, as detailed in Appendix A.[7]
- The development of a master list of "software centers," or locations where software personnel are expected to be concentrated (as listed in Appendix A). This list enabled the development of points of contact in target organizations for the data call.
- Access to workforce data to (1) cross-reference to the eventual data collected for software personnel and (2) enable the determination of whether a machine-learning algorithm could predict who performs software activities in organizations that do not respond to the data call.[8]
- Introductions to leadership within the software centers who would eventually respond to the data call.

Software Center Engagement

Through a series of engagements with targeted software centers, the research team enlisted the help of eight centers to obtain a sample of personnel who collectively perform software activities within the four categories of software identified in Chapter Three (i.e., weapon systems, warfighter support, commander support, and defense enterprise support). To identify the workforce, we used the software activity model developed by the working group. The 11 categories of software activities used to identify DoD software personnel are

- Define/Analyze Software Requirements
- Software Estimation and Planning
- Software Input to Statement of Work, Request for Proposal, and Contract Evaluation
- Software Architect/Design/Implement
- Software Code, Unit Tests, and Models/Simulations
- Algorithms, Data Analytics, and Cybersecurity
- Software Integration, Test, and Acceptance

---

[7] The purpose of the activity model is to broadly describe and categorize the universe of software activities and functions. As described in detail in Appendix A, the activity model was used to identify who performs the functions in the model and thus identify who belongs to the software workforce.

[8] The machine-learning algorithms we attempted were unable to predict who performs software activities from available data.

- Software Sustainment/Software Upgrades
- Software Processes, Documentation, Tools/Metrics
- Monitor Software Quality, Cost, and Schedule
- Programming Firmware, Logic Designer.

The detailed data collection methodology, including the selection of software entities in the sample, is included in Appendix A.

Engagement with each software center began with an introductory email sent by the project team's government sponsor representative to the commander or senior executive with ultimate responsibility for the software personnel within each software center in the target sample group. We followed up to coordinate a meeting between our data collection team and the center's leadership. The meetings occurred either in person or by telephone, depending on the availability and location of the invitees. Prior to the meeting, a read-ahead package and request for information were sent to introduce the initiative, provide background information, and provide a list of the data elements requested. During the meeting, the read-ahead materials were briefed, questions regarding the requested data were answered, and the tactics to be used by the software center personnel to collect the data were discussed. The data collection team advised the centers on how to best collect the data, and a preformatted Excel spreadsheet was provided to the action officers assigned to collect the data within the software center to standardize the data collection. The data collection team maintained ongoing communication with the software center action officers during the collection of the data to monitor status, clarify any additional questions regarding the collection, and advise the action officers as they encountered barriers when collecting the data.

After receipt of the data from each software center, the data were reviewed for validity, and any necessary follow-up with the software center occurred to make necessary corrections to obvious data errors. The data from each software center were compiled into a master spreadsheet to standardize the format of entries and then were analyzed.

Although this process may sound simple and straightforward, it should be noted that (in at least some cases) the process turned out to be much more difficult than originally thought. Even with local personnel data and local commanders supporting the initiative, the action officers often found that existing data were not easy to obtain and collecting reliable data was time consuming.[9] For these reasons, at least some of

---

[9] A common data collection challenge for the action officers was how to determine whether software was a primary duty for those identified as performing one or more software activities. To answer this question, they often needed to push the data call down to the supervisor level for input and/or review of the results. They reported difficulties in ensuring a consistent understanding of "primary duty," which analysts should bear in mind when reviewing our data.

Another challenge faced by the leadership within the software centers was timely completion of the data collection task. Only three of the responding organizations were able to collect the data in less than two weeks. Often, competing priorities arose after the data collection began. Many teams started collection only to later

the centers confined their data collection efforts to just those personnel known to be engaged in software development. Therefore, our data collection efforts do not provide a snapshot of a center's entire workforce, and the data cannot be used to answer questions regarding the prevalence of software acquisition personnel within the larger workforce.

DoD Workforce Data Call Lessons Learned and Recommendations

Early leadership buy-in is essential. The facilitation of introductions between the government sponsor leadership and the leadership of target organizations earlier in the initiative likely would have garnered additional support for the data call prior to the working group meeting. This is especially important when attempting to execute a data call for a cross-cutting functional community that does not align with the DoD chain of command.

The scope and definition of the workforce must be clear. If possible, we recommend using a reference to an existing definition that the respondent organizations are familiar with and that satisfies the scope of the data call. However, it was very clear early on that there was not a unanimous definition of "software personnel" within the software acquisition data call working group. Therefore, the research team developed and validated the simplified software activity model shown in Appendix A.

A task memo from the right leadership level in DoD would have enabled wider data collection. According to interactions with organizations involved in the data call working group and a review of other DoD workforce data calls, such as the 2018 Security Cooperation Workforce Data Call, we believe that a task memo signed at the Under Secretary of Defense level would have enabled the desired census of DoD personnel. Without such a task memo, however, it became clear that no service—even those most supportive of our efforts—would willingly engage in a census of its workforce.

In conversations with the working group and with the centers, it appears there is significant support for the idea of adding a software functional code to the DoD manpower and personnel systems and then populating the field by conducting a follow-up

---

realize they needed to collect information from multiple sources within the respondent organization to fully respond to the data call. Even at a local level, it is not always clear what personnel data is stored where or who has knowledge of workforce skill sets.

Respondents reported that the most challenging data point to collect was the number of years of software experience held. They reported that this information is not tracked in any database and could only be definitively answered by the individuals.

Note that we also initially requested the number of years of acquisition experience held. After our first four centers informed us that the information was unavailable in any database, we rescinded this request to minimize the data collection burden on individuals.

DoD-wide data call.[10] At a minimum, the data call should collect the Electronic Data Interchange Personal Identifiers and the unique position/billet ID for each position held by a person who performs software activities as defined by the software activity model in Appendix A. This would enable the identification of software personnel in the DoD manpower and personnel data systems to cross-reference against educational background, years in position, training acquired, etc. Collecting the data within the existing personnel systems would enable ongoing analysis of the workforce. Ongoing analyses, not simple snapshots of the workforce as contained in this report, will be essential to effectively target and assess workforce initiatives designed to improve the competencies of DoD's software workforce.

Most of the software centers we engaged with (even those who ultimately declined to participate) appeared eager to respond to the data call and we believe would have responded with appropriate "top cover." Because we did not obtain a task order, those who ultimately did supply data committed their own resources to collect that data, indicating a degree of commitment to our efforts. Those leaders stated that they perceived a benefit to their organization to identifying who and how many personnel were performing the various software activities. Furthermore, many leaders from the respondent organizations expressed a desire to support DoD-wide software workforce initiatives and indicated that they understood a head count is the place to start—i.e., participation in our effort was seen as a meaningful way to contribute. Many voiced an interest in comparing the data collected from their organization to that of other organizations responding to the data call, and we have provided comparisons at the service level in Appendix C.

The working group was an essential part of the data call even though it did not produce a system change requirement for the DoD manpower and personnel systems nor a coordinated data call memo. Without the engagement that occurred while organizing the data call working group and introducing the initiative to potential supporters, it is unlikely that we would have been able to collect *any* data. Among the other outcomes of the working group cited earlier, its key contribution to this research was in providing access to supporters who later made introductions to the software centers that responded to our requests for data.

Detaching a member from one of the major software centers (CCDC/Armaments Center) within the target sample population to the government sponsor's project team provided the necessary top cover via government-to-government introduction of the initiative to the sample group that ultimately participated in our study. This

---

[10]   It should be noted, however, that this support is not unanimous. We received a great deal of pushback from at least one of the services, which asserted to us that tracking and supporting its software acquisition workforce is better left to the services and that additional initiatives at the DoD level are unnecessary. Although this may be true, we were unable to obtain data to independently validate that position—i.e., we were unable to access any repositories at the service level containing the data we were seeking, either because we were not granted access or because we were granted access and found the data to be deficient.

member allowed us to maximize the utility of existing working relationships between the research team, the government sponsor representatives, and the software centers to both collect data and solicit a set of initial software centers to test and refine the instrument used in collecting that data. *We highly recommend that future software workforce researchers include one or more government personnel from the software centers as detached members of the project team.*

In the course of our research, we also gained a renewed appreciation of the fact that there are many inherently governmental tasks required to execute a functional workforce data call. Our eventual data collection efforts would not have been successful without action taken by the government sponsor. Future data calls should expect an ongoing involvement of government personnel to carry out the inherently governmental activities necessary to execute a data call of this nature. These activities include

- government sponsorship introduction (government-to-government) to the software centers and to other DoD stakeholders
- coordination of charters to organize a working group and the signing of coordinated task memos to execute a data call and other requests for information
- submission of access requests for data and system access authorizations
- sponsoring and chairing a working group
- direction of action officers within the target organization to collect the data
- activities completed by action officers and supervisors within the target organization to collect the data.

CHAPTER SIX

Summary and Recommendations

DoD has experienced persistent challenges with software development across various acquisition programs. Attempts to address those challenges through workforce initiatives, such as hiring, training, and professional development, are severely hampered by an inability to identify and characterize the presumably tens of thousands of software professionals working within DoD's acquisition programs. This report documents our attempts to identify and characterize (1) DoD's software acquisition workforce, (2) the types of software developed within DoD, and (3) the variety of methods DoD programs use to organize and train their workforce in developing mission critical software.

Although the research team gained considerable insight into the distinct challenges of DoD software and software organizations, the team was challenged in its efforts to comprehensively categorize the workforce responsible for that software. The main reason for this was a lack of data in existing DoD databases and internal DoD organizational issues that limited the team's ability to gather data. Although our research provides a snapshot of DoD's software acquisition workforce, there are limitations to this approach that did not allow us to explore how the dynamics of age, experience, and pay levels affect the hiring, promotion, and retention of DoD's software professionals. Our recommendations, which are based on lessons learned from this effort, focus on what is needed to enable future efforts to better understand, train, and employ DoD's software acquisition workforce.

The recommendations outlined in this chapter fall into three categories: those that DoD should implement immediately to improve its software development process, those that DoD should implement when delivering software training to its software workforce, and those that would allow DoD to better characterize that workforce.

DoD Software Development Process Recommendations

- It is imperative that DoD develop the competencies and culture needed to incorporate actual warfighters early and often in the development of programs and the underlying software.

- When seeking to reap the benefits of commercial software development practices, DoD must take care not to replicate the security practices of that industry.
- DoD should review current safety standards and guidance to ensure they include best practices for achieving high levels of operational trust in highly automated and/or autonomous DoD software–intensive systems.

DoD Software Workforce Development Recommendations

- DoD software workforce development initiatives should be focused on the best practices of *embedded* software development.
- Improvement initiatives should be focused on the engineering acquisition career field because most personnel in the DoD software workforce are assigned to that career field.

DoD Software Workforce Characterization Recommendations

- Additional efforts are needed to identify and characterize DoD's software workforce. Those efforts must be driven from the highest levels of the DoD hierarchy.
- Future software workforce researcher teams should include one or more government personnel from the software centers as detached members of the project team.
- DoD should conduct an in-depth analysis regarding the employment of software-educated professionals to ensure that biases regarding suitability for assignment into broader roles are not affecting promotions.
- Once the software workforce has been identified, DoD should conduct analyses to better understand how the dynamics of age, experience, and pay levels affect hiring and retention.

Considerations for Future Data Call

Our efforts under this project reinforced our belief that a full census data call of DoD's software workforce is technically possible and desirable. We continue to advocate that it be performed. However, any subsequent efforts to perform a data call would do well to consider the following barriers we encountered:

- fatigue with data calls or surveys across DoD
- inadequate resources within organizations to facilitate data collections

- lack of a perceived benefit to service acquisition executives from data calls and, for some, a belief that software should not be singled out separately from other types of acquisition
- expectations of key stakeholders that low response rates would invalidate the effort
- desires for a data collection tool other than a spreadsheet, which was felt to be error prone and inefficient, or direct entry into existing systems, which (although efficient) would not allow for independent vetting of data prior to entering it into authoritative systems
- the need for a task memo directing the data call from a level at or higher than the Under Secretary of Defense[1]
- the pursuit of competing priorities by software supervisors, which results in them having too little time to coordinate data collections in their centers.

---

[1] Other DoD-wide data calls for functional matrixed workforces (e.g., the DoD Security Cooperation Workforce) have been successful in achieving high response rates from task memos coordinated at the DoD-component-head level and then signed out at the Under Secretary of Defense level.

Data Collection Methodology

In this appendix, we describe the methodology used to collect the necessary data to characterize the DoD software workforce. Beginning with a list of DoD entities in which software personnel are known to be concentrated, we formulated a data collection methodology that considers the impacts to project schedule, existing data sources, owners of the data, and the willingness to support data collection efforts from both internal and external project stakeholders.

Workforce Composition Data Collection

Identifying DoD Software Acquisition Entities

The data collection process began with the identification of DoD entities in which software personnel are known to be concentrated. The project team developed this list through conversations with key project stakeholders, by reviewing prior research and reports related to DoD software acquisition, and through conversations with software subject-matter experts participating in the software competency development panels. These software entities include organizations colloquially referred to as DoD software centers. Table A.1 provides a list of 70 DoD entities identified as containing concentrated numbers of personnel in the software acquisition workforce. Collectively, these entities are expected to contain the vast majority of DoD's software workforce.

Identifying Software Personnel

As documented in Chapter Five, existing DoD manpower, personnel, and acquisition career management systems do not contain indicators or codes to identify who performs software activities within DoD. Furthermore, we were unable to construct a reliable algorithm using other data points in the DoD data systems to positively identify or predict who is and who is not performing software activities. The inability to identify who and how many DoD personnel perform software activities is the primary obstacle to characterizing the software acquisition workforce and is the primary driver for conducting a data call to collect the necessary data to characterize the workforce.

Table A.1
DoD Software Acquisition Entities

Organization or Command	Symbol
OSD, Joint, 4th Estate	
Missile Defense Agency/Aegis Ballistic Missile Defense System	MDA/ABMDS
Department of the Air Force	
Air Force Test Center/96th Test Wing, Cyberspace Test Group	96 TW Cyberspace Test Group
Air Force Research Laboratory, Munitions Directorate	AFRL/RW
Air Force Research Laboratory, Research Collaboration and Computing Directorate	AFRL/RC
Air Force Research Laboratory, Engineering	AFRL/EN
Air Force Research Laboratory, Information Technology	AFRL/IT
Air Force Life Cycle Management Center, Armament Directorate	AFMC/AFLCMC/EB
Air Force Life Cycle Management Center, Engineering Avionics Integration and Software Systems	AFMC/AFLCMC/EZAS
Air Force Life Cycle Management Center, Fighters and Bombers Directorate	AFMC/AFLCMC/WW
Air Force Sustainment Center HQ	AFMC/AFSC
Air Force Sustainment Center, Oklahoma City Air Logistics Complex	AFMC/AFSC/OC-ALC
Air Force Sustainment Center, Ogden Air Logistics Complex	AFMC/AFSC/OO-ALC
Air Force Sustainment Center, Warner Robins Air Logistics Complex	AFMC/AFSC/WR-ALC
114th Space Control Squadron	114 SPCS
Air Force Space Command	AFSPC
Air Force Space Command, Space and Missiles Systems Center	AFSPC/SMC
Space and Missiles Systems Center, Advanced Systems and Development Directorate	AFSPC/SMC/AD
21st Space Wing	21 SW
21st Space Wing, 21st Space Operations Squadron	21 SOPS
30th Space Wing	30 SW
Air Force Global Strike Command, 576th Flight Test Squadron	576 FTS
Air Force Test Center, Arnold Engineering Development Complex	AFTC/AEDC
Air Force Nuclear Weapons Center	AFNWC
Air Force Nuclear Weapons Center, Intercontinental Ballistic Missile Systems Directorate	AFNWC/NI
Air Force Operational Test and Evaluation Center	AFOTEC

Table A.1—Continued

Organization or Command	Symbol
Air Force Cost Analysis Agency	AFCAA
Secretary of the Air Force/Science Technology & Engineering	SAF/AQR
Secretary of the Air Force/Chief Information Officer A6	SAF/CIO A6
Air Force Software Technology Support Center (Ogden)	STSC
Department of the Army	
Combat Capabilities Development Command/Aviation and Missiles Center (formerly AMRDEC)	CCDC/Aviation and Missiles Center
Combat Capabilities Development Command/Armaments Center (formerly ARDEC)	CCDC/Armaments Center
Combat Capabilities Development Command/C5ISR Center, Software Engineering Directorate	CCDC/C5ISR/SED
Combat Capabilities Development Command/Ground Vehicle Systems Center	CDCC/GVSC
Army Research Laboratory/Army Research Office	ARL/ARO
Communications-Electronics Command/Software Engineering Center	CECOM/SEC
Communications-Electronics Command/Communication-Electronics Research, Development and Engineering Center, Aberdeen Proving Ground, Md.	CECOM/CERDEC APG
Communications-Electronics Command/Communication-Electronics Research, Development and Engineering Center, Ft. Huachuca	CECOM/CERDEC, Ft. Huachuca
Communications-Electronics Command/Communication-Electronics Research, Development and Engineering Center, Ft. Sill	CECOM/CERDEC, Ft. Sill
Communication-Electronics Research, Development and Engineering Center	CERDEC [AMRDEC]
Office of the Deputy Assistant Secretary of the Army for Cost and Economics	ODASA-CE
Army Tank Automotive Research, Development and Engineering Center	TARDEC
U.S. Army Chief Information Officer/G-6	USARMY/CIO/G-6
Army Space and Missile Defense Command	USASMDC
Department of the Navy	
Naval Surface Warfare Center Crane Division	NSWC Crane
Naval Air Warfare Center Aircraft Division	NAWCAD
Naval Air Warfare Center Weapons Division	NAWCWD
Naval Air Warfare Center Training Systems Division	NAWCTSD
Department of Navy Chief Information Officer	DON CIO
Fleet Readiness Center East	FRC East
Fleet Readiness Center Southeast	FRC Southeast

Table A.1—Continued

Organization or Command	Symbol
Fleet Readiness Center Southwest	FRC Southwest
Naval Sea Systems Command, Program Executive Office for Integrated Warfare Systems	NAVSEA/PEO IWS
Naval Center for Cost Analysis	NCCA
Norfolk Naval Shipyard	Norfolk Naval Shipyard
Naval Research Laboratory	NRL
Naval Surface Warfare Center Carderock Division	NSWC Carderock
Naval Surface Warfare Center Corona Division	NSWC Corona
Naval Surface Warfare Center Dahlgren Division	NSWC Dahlgren
Naval Surface Warfare Center Indian Head Explosive Ordnance Disposal Technology Division	NSWC IHEODTD
Naval Surface Warfare Center Panama City Division	NSWC PCD
Naval Surface Warfare Center Dahlgren Division Dam Neck Activity	NSWCDD Dam Neck Activity
Naval Surface Warfare Center Philadelphia Division	NSWCPD
Naval Surface Warfare Center Port Hueneme Division	NSWCPHD
Naval Undersea Warfare Center Keyport Division	NUWC Division Keyport
Naval Undersea Warfare Center Newport Division	NUWC Division Newport
Space and Naval Warfare Systems Command	SPAWAR
Space and Naval Warfare Systems Command Headquarters	SPAWAR HQ
SPAWAR Systems Center Atlantic	SSC Atlantic
Marines Corps Combat Development Command	MCCDC
Marines Corps Tactical Systems Support Activity	MCCDC/MCTSSA

SOURCE: Country Intelligence Group Analysis.
NOTE: The actual number and type of software acquisition personnel is not known for all entities listed in this table. The list was developed to be as inclusive as possible and to provide a list of the probable organizations within the universe of DoD software acquisition personnel. The list also contains some overlap between an organization's headquarters or major command and the local offices or subordinate commands within the organizational hierarchy to enable the geographic mapping of software entities (i.e., the headquarters may have oversight of all software personnel in its hierarchy, but the personnel are geographically dispersed at non-headquarters locations).

Software Activity Model

We engaged stakeholders within OUSD(R&E) and other software acquisition subject-matter experts during a series of working group meetings to develop a list of software functions that, if performed, would indicate the person is involved in software acquisition, development, or sustainment. The following 11 categories of software activities were used to identify DoD software personnel:

- Define/Analyze Software Requirements
- Software Estimation and Planning
- Software Input to Statement of Work, Request for Proposal, and Contract Evaluation
- Software Architect/Design/Implement
- Software Code, Unit Tests, and Models/Simulations
- Algorithms, Data Analytics, and Cybersecurity
- Software Integration, Test, and Acceptance
- Software Sustainment/Software Upgrades
- Software Processes, Documentation, Tools/Metrics
- Monitor Software Quality, Cost, and Schedule
- Programming Firmware, Logic Designer.

The 11 categories were also tested to determine how well they perform as a model to identify software personnel. During working group discussions, the Air Force Life Cycle Management Center (AFLCMC) Engineering Directorate offered to help test the validity of the software activity model within the AFLCMC Avionics Software Engineering and Integration (EZAS) office. The EZAS Chief delivered the activity model to all 43 personnel in their office using Microsoft Outlook voting buttons requesting that personnel answer "Yes" if they perform at least one of the software activities and "No" if they perform none of the software activities in the model. Figure A.1 shows the image of the software activity model that was delivered to the personnel as an aid to help them respond accurately.

We received 37 responses within two days, and 27 personnel in the EZAS office responded that they perform at least one of the software activities in the model. The EZAS Chief reviewed the results compared with the expected answer (according to their leadership's familiarity with the tasks and duties assigned to the individuals) and determined that six of the personnel answered "No" when they were expected to answer "Yes." All personnel who were expected to answer "No" did so.

Testing the model within EZAS showed that when the model was delivered via a survey (asking people to self-identify), it accurately identified 81 percent of the personnel who perform software activities. But it accurately excluded 100 percent of personnel who do not perform software activities. Following the test within EZAS, no changes were recommended to the model or the software activities within it, but we

Figure A.1
Software Activity Model

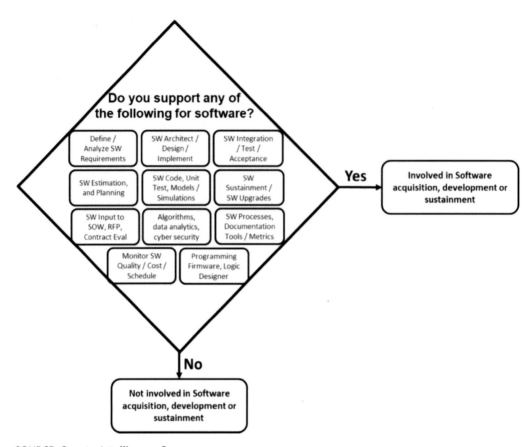

SOURCE: Country Intelligence Group.
NOTE: RFP = request for proposal; SOW = statement of work; SW = software.

recommended that office chiefs and supervisors with familiarity of their employee's tasks and duties assigned should respond on behalf of their employees rather than using a survey asking personnel to self-identify.

Challenges with Conducting a DoD-Wide Software Acquisition Personnel Data Call

As described in Chapter Five, several challenges arose during the planning and implementation phases of the intended data call that prevented a DoD-wide data call to identify all software personnel in DoD. First, the organizational realignment of the former OUSD(AT&L) occurred right as the data call was being planned. This resulted in difficulty identifying the appropriate responsible parties for forming an effective data call working group and changes in support services provided to the successor organizations of the OUSD(R&E) and the OUSD(A&S). The latter resulted in the inability to develop an automated data collection instrument within the confines of the existing support structure available to the OUSD(R&E) after the reorganization.

Second, there were significant changes to leadership within the government sponsor's office and vacated positions that would be necessary to engage the service acquisition executives and DACM offices on a government-to-government basis to coordinate the drafting and concurrence of a task memo to complete the data call.

Third, there was general resistance to conducting a DoD-wide data call among the DACMs because they were not convinced of the benefit of conducting such a data call for software acquisition.

Finally, there was a strategic pause on DoD career field FIPT activity that also hindered the ability to engage functional leaders and gain their concurrence for a DoD-wide data call.

Because of the combined impact of the challenges encountered, a DoD-wide data call was not possible within a reasonable time frame to support the completion of the study. Therefore, data collection efforts were focused on collecting a representative sample of the workforce by directly engaging program managers and commanders within individual software centers. Note that although there was some overlap between the software centers we collected workforce characterization data from and the program offices that participated in our workforce employment interviews, the two efforts were largely independent.

Sample Selection

We emphasized achieving a representative sample of the software workforce during the selection of entities to include in the target sample group. The sample was also shaped by the project team's and government sponsor's perceptions of which centers might support and respond to a data call. The initial list of organizations included ten entities:

- AFLCMC/EZAS
- Air Force Research Laboratory, Munitions Directorate (AFRL/RW)
- Air Force Life Cycle Management Center, Armament Directorate (AFLCMC/EB)
- Naval Air Warfare Center Aircraft Division (NAWCAD)
- Naval Air Warfare Center Weapons Division (NAWCWD)
- Air Force Test Center/96th Test Wing (96 TW), Cyberspace Test Group
- Air Force Life Cycle Management Center, Fighters and Bombers Directorate
- Naval Surface Warfare Center Crane Division (NSWC Crane)
- Army CCDC/Aviation and Missiles Center
- Army CCDC/Armaments Center.

Additional software centers were added to the target list to ensure a greater balance across the military departments, to ensure the inclusion of offices of varying size, to consider inclusion of personnel occupying both acquisition-coded and nonacquisition-coded positions, and to ensure the inclusion of personnel supporting software for a variety of systems, including weapon system platforms, vehicles, battle management systems, armaments, communications, and business enterprise systems.

Tables A.2 and A.3 show the expanded list of all 17 entities invited for inclusion in the sample group. Eight entities provided data resulting in a sample size (*n*) of 2,123 software personnel that included personnel in both acquisition- and nonacquisition-coded positions within all military departments and selected agencies, supporting a variety of systems.

Outreach and Request for Information

Each entity invited for inclusion in the sample received an introductory email from the government sponsor's representative requesting a meeting to brief the entity's leadership on the study objectives and to explain the need to collect the requested data. Data for the following variables related to personnel information were included in the request for information:

- educational background (degrees obtained and major area of study)
- personnel type (uniformed or nonuniformed)

Table A.2
DoD Software Centers That Supplied Characterization Data

Organization or Command	Symbol	Number of Software Personnel	Number of Personnel at Location
Air Force Research Laboratory, Munitions Directorate	AFRL/RW	242	305
Air Force Research Laboratory, Research Collaboration and Computing Directorate	AFRL/RC	9	71
Air Force Life Cycle Management Center, Armament Directorate	AFMC/AFLCMC/EB	20	541
Air Force Life Cycle Management Center, Engineering Avionics Integration and Software Systems	AFMC/AFLCMC/EZAS	37	43
Naval Air Warfare Center Aircraft Division	NAWCAD	641	11,912[a]
Naval Air Warfare Center Weapons Division	NAWCWD	390	9,037[a]
Army Combat Capabilities Development Command/Armaments Center	CCDC/Armaments	741	3,380
Missile Defense Agency/Aegis Program, Ballistic Missile Defense System	MDA/ABMDS	43	210
Total		**2,123**	**25,572**

[a] Estimate uses total uniformed and nonuniformed DoD personnel at location per DMDC Reporting Service.

Table A.3
DoD Software Centers Invited to Participate but That Ultimately Were Unable to Supply Characterization Data

Organization or Command	Symbol
Air Force Test Center/96th Test Wing, Cyberspace Test Group	96 TW Cyberspace Test Group
Air Force Life Cycle Management Center, Fighters and Bombers Directorate	AFMC/AFLCM/WW
Air Force Life Cycle Management Center, Fighters and Bombers Directorate, F-16 System Program Office	AFMC/AFLCMC/WWM
Air Force Sustainment Center, Oklahoma City Air Logistics Complex	AFMC/AFSC/OC-ALC
Air Force Sustainment Center, Ogden Air Logistics Complex	AFMC/AFSC/OO-ALC
Naval Surface Warfare Center Crane Division	NSWC Crane
Combat Capabilities Development Command/Aviation and Missiles Center (formerly AMRDEC)	CCDC/Aviation and Missiles Center
Combat Capabilities Development Command/C5ISR Center, Software Engineering Directorate	CCDC/C5ISR/SED
Communications-Electronics Command/Software Engineering Center	CECOM/SEC

- rank or grade
- nonuniformed occupational series or uniformed occupational specialty, Air Force specialty code, or branch code
- occupying an acquisition-coded position
- years of acquisition experience (if in acquisition-coded position)[1]
- acquisition certification level required (if in acquisition-coded position)
- acquisition certification level achieved (if in acquisition-coded position)
- years of software experience
- time in current position.[2]

This information was expected to be contained within existing systems from which the software center could readily extract the data. In addition, the following software activity information was requested:

- is software the position's primary duty (i.e., software activity is critical to the position's duties, role, or responsibilities)?
- software activity categories performed (from the activity model)

---

[1] Years of acquisition experience was removed after analyzing the data from the first four entities because we found the data were either not obtainable in existing systems or contained invalid or incorrect entries.

[2] Time in current position was removed after analyzing the data from the first four entities providing data because we found the data was either not obtainable in existing systems or contained invalid or incorrect entries.

- list of software activities that are central to the job role (if any)
- position descriptions.

Our data collection team prepared the briefing materials and coordinated a meeting with the contacted entity. Meetings occurred either in person or by telephone, depending on the entity's preference. The briefing consisted of a project summary and status, an explanation of the data points requested, the provision of a preformatted data collection spreadsheet to improve the normalization of data received, and a discussion about the data sources and the recommended techniques to collect the data. Using the results of the AFLCMC/EZAS data collection test, we recommended that supervisors with familiarity of the tasks and duties assigned should respond on behalf of personnel within their organization rather than executing a survey requesting personnel to self-identify that they perform software activities. All respondent organizations except for AFLCMC/EB and the Aegis Ballistic Missile Defense System (ABMDS) Program Office tasked supervisors or an action officer with familiarity of the task performed to collect and respond to the data request. AFLCMC/EB and the ABMDS Program Office opted to survey their personnel because they did not readily have access to all the personnel data requested.

Data Collection and Compilation

Most entities that responded to the request for information provided the data within 30 days following the initial meeting with the organization's leadership. All responding organizations provided data using the preformatted spreadsheet. We reviewed the data for valid entries and followed up with the respondent organizations as necessary to clarify entries. Each submission was compiled into a master spreadsheet and was cross-referenced with additional data sources provided from working group participants to fill in blank data element entries for matching records. Additional sources of information were also cross-referenced with the data received to provide a summary of the total number of personnel reported as performing software activities compared with the estimated total personnel within each respondent organization as shown in Table A.2.

Other Sources of Bias

We identified two sources of bias that could affect the results of the data collection. The first is the use of a survey by AFLCMC/EB and the ABMDS offices allowing personnel to self-identify if they belong in the sample of software personnel. This may have resulted in underreporting of those performing software activities within those organizations and could affect the objectivity of the answer to the question concerning whether software is a primary duty or whether software competencies are central to their job role.

The other potential source of bias is present simply because the respondent organizations consist of those who were willing to respond without a directive or formal task memo. If a respondent organization had a vested interest in the outcome of this study, the small number of samples collected would give them ample opportunity to inject bias. It should be noted, however, that the data collection team did not detect any intent to bias the results during discussions with the respondent organizations, and all who responded appeared to do so objectively.

Workforce Employment Data Collection

Selection of Participants for Interviews

As noted in Chapter Four, we originally planned to interview a representative sample of program offices as determined by the distribution of software value across the FY 2018 major programs (shown in Figure A.2).

Unfortunately, we were unable to secure a task memo from OSD or from the service acquisition offices to facilitate our interview requests and had difficulty securing the cooperation needed to achieve a representative sample. Instead, we solicited participation using the personal contacts of the sponsor and of our research team to perform a snowball sample. Using this methodology, we were able to obtain interviews across the DoD services and for both embedded and enterprise systems, but we caution against reading too much into the observations we developed from the interviews. As noted in Chapter Four, the sample is heavily weighted toward Air Force and enterprise software and underweighted for the Navy and embedded systems that dominate software acquisitions within DoD.

Interview Protocol

The interview protocol used had two versions, with the second including questions regarding difficulties in obtaining training. These questions were added after our initial two interviews indicated that some programs were encountering significant barriers obtaining training in modern software development techniques and were having to learn on the job. The added questions are highlighted using italics in the protocol shown in Box A.1.

Sources of Bias

Our interviews were weighted toward Air Force, Joint, and enterprise systems, as shown in Table 4.1. We highly recommend that future characterizations of the DoD software workforce interview a more representative sample to improve our understanding of challenges faced by program offices when employing their software workforce. Note that because of the lack of a representative sample, we do not derive findings or recommendations from the program office interviews.

Figure A.2
Recommended Process to Draw Samples for Program Interviews

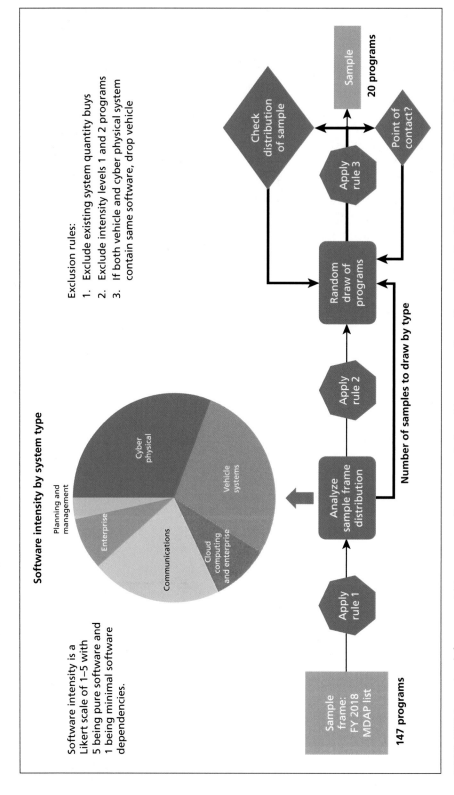

NOTE: MDAP = major defense acquisition program. Process was designed to preserve the software intensity distribution of the sample frame (the FY 2018 major defense acquisition program list).

Box A.1
Program Manager Interview Protocol

Introduction

Thank you for speaking with us today. [As I mentioned in my email,] OSD AT&L(SE) asked RAND NDRI to help characterize the software acquisition workforce within the DoD and to collect data regarding how programs employ that workforce. As part of our study, we are attempting to gather data regarding how programs organize, staff and manage their software acquisitions.

During our discussion, we will take notes so we can accurately capture your comments for our analyses and report. In our report, we will broadly describe the types of experts represented in these discussions, but we will not identify you by name or title and we will not attribute quotes to you or your organization. Even so, we acknowledge that if your perspective or experiences are very unique, it is possible that informed readers will be able to infer your identity.

Your participation in this study is completely voluntary. You may choose not to participate, or not to answer any question at any time. The discussion should take about 45 to 60 minutes. Do you have any questions before we begin?

We would like to start with some general information about the type of software being acquired for your program so we can better tailor our follow-on questions.

Background Questions
1. Can you briefly identify the major software elements of your program and describe the role of each within the overall program?
2. In your view, how reliant is the program on the software being acquired? [Alternatively: In your view, which elements of the software are the most critical to program success?]
3. Cost and Schedule
   a. Are any of the software elements currently on the program critical path?
   b. Can you provide an order of magnitude estimate of the cost of the software being acquired (relative to the total program costs)?
4. Is the program utilizing agile system or software development, deployment or delivery methods for this acquisition? Can you describe the methods being used?

Questions on Staffing
1. Roughly, how large is your total acquisition workforce for this program?
2. What percentage of their time or effort is devoted to acquisition of software?
3. Does the program have a sufficient number of trained software acquisition professionals? [If answer is no], what skill sets or competencies are particularly lacking?
4. How do the software professionals on your team keep up to date with modern software development methods, processes and tools?
5. What are the barriers to keeping software competency current and up to date?

Questions of Organization
1. Does the program have a software Chief Engineer or Architect? Can you describe their role?
2. Who on the program acquisition staff serves as the point person if/when a critical software issue occurs?
3. How are Integrated Product Teams (or other teaming construct) organized for this acquisition?
4. What is the reporting path if any of those teams experience critical issues with software acquisition?
5. How are issues resolved that arise between teams? [particularly for interfaces between (a) software and system teams and (b) software and hardware teams]

Questions of Employment
1. Are any program acquisition staff embedded within contractor system engineering or software teams?
2. Are any program acquisition staff designated as "the voice of the customer" when software teams have questions regarding how the software should function or perform? [If yes] is that assignment a formally recognized role or is it more ad hoc? [If no] how do software teams resolve questions regarding software function or performance?
3. Who validates the software requirements for this program? [i.e. who determines if the requirements accurately reflect the goals and objectives of the program]

Box A.1—Continued

4. Who verifies the software requirements for this program? [i.e. who determines that the delivered software meets the goals and objectives of the program]
5. Are any program acquisition staff designated as "the voice of the customer" when software teams have questions regarding how the software should be delivered into the test or operations environment? [If yes] is that assignment a formally recognized role or is it more ad hoc? [If no] how do software teams resolve questions regarding delivery into test or operations?

Are there any questions we should have asked you, but didn't?

Closing Remarks
Thank you so much for meeting with us today. The information you provided is critical for our analysis. Our study will be completed later this year at which time we will be providing our project sponsor a final briefing and eventually a published report that we would be happy to share with you.

Workforce Data Analysis

In this appendix, we provide technical detail on the construction and characterization of the DoD software workforce sample used in our characterization analyses. This is followed by a section detailing our analyses comparing the DoD software workforce with industry. In this second section, we have included additional comparative analyses that are representative of the type of analyses we believe DoD will need to conduct if it is to meet its goals of understanding the challenges of recruiting and retaining this workforce. We chose not to include these analyses in the main body of the report, given the extensive assumptions needed to conduct them.

Construction and Characterization of the DoD Sample Used in Our Analyses

The research team's request for information collected survey results from 2,123 individuals across eight organizations within the Army, Navy, Air Force, and OSD.[1] Table B.1 provides the number of individuals from each organization and the percentage of the total.

One of the first challenges we faced in making sense of the data was that some offices included their entire staff, while others (because of resource constraints) needed to limit their data collection to those personnel engaged in software development activities. Ideally, we would have liked all organizations to have reported all staff. This would, perhaps, provide insight into questions regarding the composition of staff engaged in software activities versus those not engaged. We determined that the quality and quantity of our data would not support analysis of that type. Having made that determination, we then removed from the sample those individuals not engaged in software activities.

---

[1] In our discussion of the characteristics of our sample by service in Appendix C, the OSD data (Missile Defense Agency/Aegis Program) were combined with Navy data. We were concerned that reporting these data separately might compromise the anonymity of the personnel reported by this office.

Table B.1
Number of Responses, by Software Acquisition Entity

Organization	Number of Responses	Percentage of Responses
AFLCMC	57	3
AFRL	251	12
CCDC	741	36
NAWCAD	641	31
NAWCWD	390	1
Missile Defense Agency/Aegis Ballistic Missile Defense	43	2
Total	**2,123**	**100**

NOTE: Percentages are rounded to the nearest whole number. Two organizations are included in the totals for AFLCMC (EB and EZAS) and AFRL (Munitions Directorate and Research Collaboration and Computing Directorate), making eight total organizations that contributed data.

Initially, we considered eliminating all individuals who answered "No" to the question of whether they identified as being part of the software acquisition workforce. Unfortunately, 750 records did not contain a response to this question, and it was clear from responses to other questions (such as whether software was a primary activity or those who listed the software activities performed) that many of these individuals were performing software activities. Ultimately, we eliminated from the sample those individuals who meet all of the following three criteria:

1. the answer to whether they identify as part of the software acquisition workforce is "No" or blank
2. the answer to whether software-related work is their primary duty is "No" or blank
3. their software role was "None" or could not be identified.

This leaves a sample of 1,911 personnel who we use in our analyses. Table B.2 shows the breakdown by organization of the 10 percent of personnel excluded from the sample by this process and the resulting percentage within the sample from each organization.

The personnel in this sample are 1 percent uniformed enlisted or officers. The · vast majority (99 percent) are career nonuniformed personnel. All uniformed personnel are Air Force personnel. According to their acquisition career field, 78 percent of personnel are coded in engineering, 8 percent in science and technology, and 8 percent are in system engineering or testing and evaluation. As shown in Figure 2.1, only 1 percent are coded as belonging to the IT acquisition career field.

Table B.2
Number of Software Personnel by Software Acquisition Entity

Organization	Number of Records Collected	Number Excluded as Not Software Personnel	Remaining Number of Records	Percentage of Sample
AFLCMC	57	4	53	3
AFRL	251	72	179	9
CCDC	741	136	605	32
NAWCAD	641	0	641	36
NAWCWD	390	0	390	20
Missile Defense Agency/Aegis Ballistic Missile Defense	43	0	43	2
Total	**2,123**	**212**	**1,911**	**100**

NOTE: Percentages are rounded to the nearest whole number.

Individuals within the sample fell into a variety of pay plans, including the General Service pay scale and the DoD Civilian Acquisition Workforce Personnel Demonstration Project (hereafter referred to as the "ACQ Demo"): Demonstration Professional, Demonstration Air Force Scientist and Engineer, and Military Grades.[2] For comparison in certain analyses, all pay grades within each plan were converted to their ACQ Demo equivalent (NH). These broke down into NH-02 (20 percent), NH-03 (60 percent), NH-04 (20 percent), and senior-level staff (0.2 percent).

In terms of the sample's terminal education degrees, 55 percent held bachelor's degrees, 39 percent held master's degrees, and 5 percent held doctoral degrees. Using the major area of study within these degrees, we found that 51 percent of individuals had educational backgrounds directly related to software and/or computer science, and 45 percent had degrees or certifications in another STEM field.[3] Figure 2.2 provides the breakdown by educational background.

To understand how organizations employ their software workforce, we asked for position description, the primacy of software in their normal duties, and which activities they performed within a software acquisition workflow. As reported earlier, this information was not readily available in any database and took significant resources

---

[2]   See DoD, *DoD Civilian Acquisition Workforce Personnel Demonstration Project (AcqDemo) Operating Guide*, Version 3, March 8, 2019, for more detail on these pay bands and their relationship to the GS scale.

[3]   A degree in software or computer science at any level (bachelor's, master's, or doctoral) was counted as educational background. For those without a software-related degree, we used the terminal degree in classifying their educational background.

to collect. Not all offices were able to respond to all questions.[4] In particular, we were told that collecting data regarding whether software constituted a primary duty was difficult to obtain: 42 percent responded "Yes," 24 percent said "No," and 34 percent of the records were blank.

For software activities performed, selections were made from the categories shown in Figure A.1's software activity model, such as "Define or analyze software requirements," "[Perform] software coding, unit testing, modeling, and simulation," or "Provide input to statements of work, requests for proposals, contract evaluation, and award-related to software." Using these responses, we categorized individuals into three software roles: Developer (41 percent), SEIT (17 percent), and Manager (6 percent). Individuals who did not answer this question were classified as Unknown (37 percent).[5] This classification was by process of elimination: Individuals who performed software architecture, design, implementation, or unit test were classified as Developers; individuals who did not perform those activities but did perform system engineering and integration activities (such as requirements analysis, integration and test) were classified as SEIT; the remaining individuals who only performed planning and oversight tasks were classified as Managers. Therefore, a manager is not a position but a role—i.e., a hands-on manager who performs all software activities, including design and implementation, is characterized as a Developer using this classification scheme.

Table B.3 categorizes our sample by educational background and software role. For completeness, Table B.4 shows a cross tabulation of the 10 percent of individuals excluded from the sample. As expected, the exclusion removed all personnel who responded that they did not perform any software activities (software role = "None"). It also excluded 26 personnel with an educational background in software and 140 individuals overall who answered no or did not answer any of the questions that would have allowed us to positively identify them as software personnel (i.e., the activity workflow, "do you self-identify as a member of the workforce?" or "is software your primary duty?"). The largest percentage of individuals removed from the sample (3.1 percent) have an educational background in "other engineering," but no single educational background was disproportionally affected.

---

[4]   Null values were clustered by organization. NAWCAD did not supply data for any of the software-related questions. AFLCMC/EZA provided data as to whether software constituted a "primary duty," but did not elaborate in subsequent software-related questions. Also note our prior caution that achieving a uniform definition of primary duty was one of the difficulties the action officers reported.

[5]   Individuals who answered "Does not perform any software function" were removed from the sample during our initial screening for personnel who are not engaged in software activities.

Table B.3
Number and Percentage of Individuals in DoD Software Personnel Sample, by Educational Background and Software Role

Educational Background	Software Role					Total	Percentage
	Developer	SEIT	Manager	None	Unknown		
Software-related	486	162	23	0	273	944	49
Aerospace and electrical	129	63	28	0	375	595	31
Other engineering	91	44	29	0	3	167	9
Other STEM	41	18	12	0	5	76	4
Non-STEM	23	36	16	0	8	83	4
Unknown	5	4	3	0	34	46	2
Total	775	327	111	0	698	1,911	
Percentage	41	17	6	0	37		100

NOTE: Percentages are rounded to the nearest whole number and may not sum to total shown.

Table B.4
Number and Percentage of Individuals Excluded from DoD Software Personnel Sample, by Educational Background

Educational Background	Software Role		Total	Percentage Removed
	None	Unknown		
Software-related	6	26	32	2
Aerospace and electrical	24	30	54	3
Other engineering	20	46	66	3
Other STEM	12	7	19	1
Non-STEM	9	24	33	2
Unknown	1	7	8	0
Total	72	140	212	10

NOTE: Percentages are rounded to the nearest whole number and may not sum to total shown.

Quantitative Analyses Comparing Software Workforce Data from DoD and Industry

On a general note about the empirical analyses in this report, the limitations of the underlying DoD workforce sample data are extensive. For many of the analyses, we needed to make assumptions, requiring varying degrees of speculation, to make progress on the questions we were tasked with studying. In this appendix, we endeavor to clearly state the extent of this issue and to present appropriate caveats in the narration of the analyses.

In addition to the DoD sample, we use data on the U.S. workforce and a subset of workers in the private computer industry for various comparisons. These data are from the ACS, an annual 1 percent survey of the U.S. population that collects a variety of demographic and labor market data.[6] We use ACS microdata (one-year estimates) from 2016 to 2018. We attempted to more closely match the characteristics of the DoD workforce sample by restricting our analyses using these ACS data to individuals with an associate's degree or higher level of educational attainment and by restricting the sample to individuals aged 21 and older and with 50 years or less of experience (the construction of this measure is detailed in the following section). There are two significant disconnects between these two samples that hamper direct comparisons. The first that we will discuss is the ability to equate age in the industry sample with years of experience in the DoD sample. The second is the ability to equate pay scales in the DoD sample with salary in the industry sample.

Construction and Analysis of Experience

The experience measure we constructed for the DoD software workforce sample comprises the answer to the question about "years of software experience"; or, where these data were missing, "How long in current position?" was substituted, if present. Recall that we dropped the question regarding how long the individual has been in their current position after noticing that respondents were having difficulties obtaining this data. Although it was also difficult to collect "years of software experience," we retained that question as essential to our research. For those who answered both questions, "years in current position" and "years of software experience" are perfectly colinear for the vast majority. However, for the one center that differentiated between these questions, our data show that most workers brought significant software experience into their current position, there were relative few who had hired into the position (i.e., the values were equal), and only one had gained their software experience after having

---

[6]   Many economists may be wondering why we did not use the extensive data available from the U.S. Bureau of Labor Statistics and Census websites as the industry basis for our comparative analysis. In fact, we attempted to use these data for some of our analyses. Doing so required us to make additional assumptions regarding how our collected data could be mapped against the industry data. Our confidence in these assumptions was low enough that we elected not to use the results as a basis for making recommendations to DoD.

been in the position for several years. Therefore, we believe that the experience measure constructed using our method is a valid measure of actual software experience.

Unfortunately, the same cannot be said for the measure we constructed for the ACS sample. Experience for the ACS sample was derived using an approximation approach common in labor economics studies. In this instance, we estimated experience as age minus 13 years of primary and secondary education minus additional years of education based on postsecondary educational attainment (with an associate's degree coded as two years, a bachelor's degree coded as five years, a master's degree coded as seven years, and a Ph.D. coded as ten years). For a small number of individuals who were assigned negative experience through this approach, we assigned zero years. We also dropped a small number of individuals who were assigned more than 50 years of experience. In equating this age-based metric to experience, we made the implicit assumption that those working in the software industry have done so since the beginning of their careers. This is a problematic assumption, given the rapid growth of the software profession. A quick search of online forums produces stories from many developers who say they began their software careers in their 30s.[7] We searched the literature for other instances in which researchers asked specifically for years of software experience. One 2015 paper, with an admittedly small sample size (about 400 people recruited for the study during a one day "code retreat"), has an experience profile a bit like our sample, in that there is a significant drop in the number of people with more than ten years of experience.[8]

Given that our industry profile is clearly age-based and not experience-based, the question is whether that matters to our analysis. We used the experience profile in two ways. The first is a qualitative check as to whether the dynamics within DoD appear to match that in industry, which they did not. From this result, we conclude that more research is required and that the dynamics seen in DoD are not those we would expect of a profession that is relatively static and stable in terms of entry and exit. This is a valid observation.[9] The second way in which we used the years of experience is in generating the salary differentials. If experience is a stronger determinate of salary than age is in the industry sample, then the higher salaries in industry that we found at higher years of experience are understated. If age is more important than experience in setting

---

[7]   Search items included "is 30 really too old to start a career as a developer?" which has a lively Quora discussion.

[8]   See Figure 1 in David Parsons, Teo Susnjak, and Anuradha Mathrani, "The Software Developer Cycle: Career Demographics and the Market Clock: Or, Is SQL the New COBOL?" *Proceedings of the ASWEC 2015 24th Australian Software Engineering Conference*, Vol. II, Adelaide, Australia, September 28–October 1, 2015, p. 87.

[9]   Note that an *invalid* observation would be that the software industry as a whole does not share these dynamics.

DoD salaries, then the comparisons may be valid.[10] We will return to this topic in the following sections.

Figure B.1 shows the distribution of years of software experience for the DoD sample. Figure B.2 shows two curves related to the experience distribution of the industry workforce. The first, for a broader comparison, shows the experience of all employed individuals in the ACS (subject to the other sample restrictions discussed above). The second curve is the subset of these individuals working in the "Computer systems design and related services" industry (Census code 7380). For this three-year period, this subsample has around 70,000 workers in it.

As seen in Figure B.2, our industry data show that software personnel are relatively young compared with the general workforce, a characteristic shared by the DoD sample (which shows most workers have relatively few years of software experience). However, the industry software personnel experience profile shows none of the dynamisms of the DoD profile. We suspect this is due to late career entrants smoothing out the actual experience dynamics. Given the significant differences between the profiles of the DoD software and industry profiles, we make the following observations:

- DoD's software workforce experience profile is not likely to be age-based.
- It is more likely that the experience profile is being driven by (1) factors that are inherent to a young industry (such as software), (2) distinct DoD hiring and retention policies, and/or (3) the small sample size or distinct characteristics of the DoD software centers we collected data from.

Construction and Analyses of Salaries

Among the individuals in the DoD sample, salary data were not collected directly, but (as mentioned earlier) data on the pay band of each individual were collected. For the small number of active duty officers in the sample, uniformed personnel pay grades were crosswalked to the NH scale. All alternate pay bands were then crosswalked back to the General Schedule (GS) scale, and a salary was imputed for each individual by taking the average of each associated GS classification using the national table (i.e., not location specific) and averaging across the amounts associated with all the years of experience in each GS pay band. For the ACQ Demo pay bands, which generally each comprise multiple GS scale steps, the average of these averages was taken.[11] For ACS individuals, wage or salary income is asked directly.

Although we could have split the ACS salaries into bands and averaged them to provide a more apples-to-apples comparison, we do not believe that extra step is

---

[10]   Note that both DoD and industry salaries ideally are merit-based. Although neither age nor years of experience necessarily imply greater competency, it is true that years of experience is a closer approximate of competency than age is (assuming we learn from our experience).

[11]   Overall, the sample used for salary analyses is a bit smaller than the full DoD sample (N = 1,911), since some individuals were missing a usable pay band entry (39 individuals).

Figure B.1
Workforce Experience Profiles in DoD Sample

SOURCE: RAND analysis of collected DoD software personnel data and of U.S. industry workforce data derived from ACS microdata from 2016 to 2018.

Figure B.2
Workforce Experience Profiles in Industry Sample

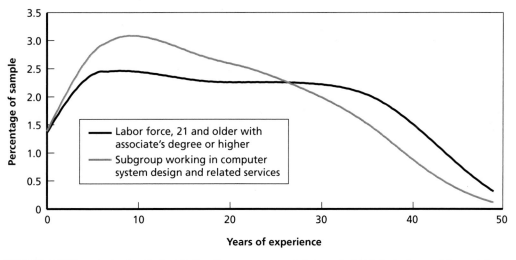

SOURCE: RAND analysis of collected DoD software personnel data and of U.S. industry workforce data derived from ACS microdata from 2016 to 2018.

needed. Although the averaging of these pay bands will introduce some error, since (by definition) everyone is assigned the mean value of their pay group, and no one is assigned to the tail values of these distributions, this error does not meaningfully affect any of the analyses we conducted. Our analyses consider only group-level averages and thus are valid within that constraint.[12]

One of the first analyses we performed that relied on wage and salary data was a comparison of DoD versus industry pay by years of experience/age as a function of educational background. Fortunately, educational background could be derived identically for both sets of data. Figures B.3 and B.4 show the results of this analysis.

For those with a STEM degree, DoD salaries start at a comparable level with industry pay but do not keep pace as experience increases. For those with a STEM degree and more than ten years of experience, we see estimated salary gaps of $25,000 to $45,000 per year relative to private industry compensation. A different pattern is observed for non-STEM degrees. In this case, there appears to be a premium in initial DoD pay of around $25,000 per year. Those with high levels of experience see a salary gap, but it is lower than that seen by professionals with STEM degrees (i.e., around $15,000 per year). These differentials are plotted explicitly in Figure 2.7.

However, for the differentials to be valid, we need to assume that age in our industry data is a reasonable proxy for years of experience in the DoD sample. As we discussed earlier, this may not be a valid assumption, given the relative youth of the software profession. If experience is a stronger determinate than age of salary in the industry sample, then the higher salaries in industry that we found at higher years of experience are understated. If age is more important than experience in DoD salaries, then the comparisons may be valid. Note, however, that both DoD and industry salaries ideally would be merit-based. Although neither age nor years of experience necessarily imply greater competency, it is true that years of experience is a closer approximate of competency than age (assuming we learn from our experience). In actuality, both the industry and DoD samples show a strong correlation of salary to years of experience (for DoD) and to age (for industry), with one exception: DoD software personnel with a non-STEM background. This observation led us to examine this group of personnel more closely and revealed that about half of the late-career software professionals within DoD have a non-STEM background. Not only are they paid better

---

[12]   To make this assertion concrete, we can compare the mean and standard deviation of our actual assigned salary averages, which use the mean of each pay band, with an alternate approach that randomly assigns tenure positions within each step on the GS scale for those who provided GS-based pay scale information and that randomly assigns both GS-equivalent grades and tenure positions using the provided NH-based pay scale information for those not on the GS scale. Using the averaging method to estimate salaries, the mean value in the sample is $88,741, and the standard deviation is $22,742. Randomly assigning values of tenure and grade and tenure as described earlier (using a uniform distribution, which assigns an equal chance of each combination of these values occurring), the mean salary level in the sample is $83,934, and the standard deviation is $21,946.

Our methodology does not support analyses on differences in salary within a pay band (for example, workers in the NH-03 pay band).

Figure B.3
Annual Salary Levels, by Experience/Age and Educational Background (DoD Sample)

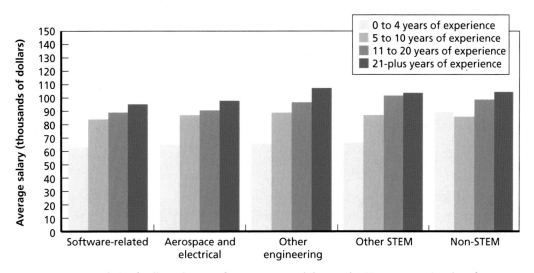

SOURCE: RAND analysis of collected DoD software personnel data and ACS one-year microdata from 2016 to 2018.

Figure B.4
Annual Salary Levels, by Experience/Age and Educational Background (Industry Sample)

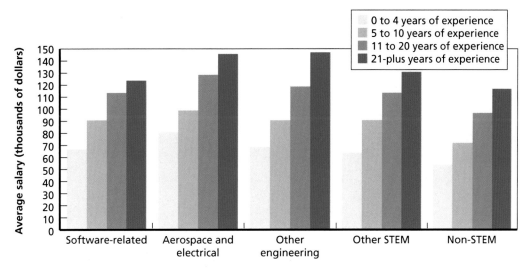

SOURCE: RAND analysis of collected DoD software personnel data and ACS one-year microdata from 2016 to 2018.

than they might be in industry, their pay appears to be independent of years of experience. A more complete data set, such as what we could achieve with DMDC data for this group of individuals, might answer whether these observations are a result of highly competent individuals being hired into DoD late in their careers versus a result of higher salaries inducing them to stay in DoD rather than taking their chances in the commercial sector.

Additional Analyses Comparing DoD to Industry

Salary Comparisons by Occupational Grouping

We also attempted to compare wage or salary between DoD and industry by software role. This comparison required that we find a method to derive relatively plausible counterparts to the roles of developer, SEIT, and manager within the industry labor market. This was not straightforward and ultimately required that we instead find three groups in the labor market that we could map back to the DoD sample using both software role and acquisition career field. Table B.5 presents a matrix of these two fields from the survey and shows the mapping from these cells to the occupational U.S. Census codes for (1) computer programmers, software developers, and computer and information research scientists (in green), (2) software quality assurance managers and testers (in orange), and (3) computer and information systems managers and engineering managers (in yellow). The occupational groupings of interest thus broadly encompass personnel who appear to be most directly involved in the development of software; those who are involved with the manufacturing, quality assurance, and testing of software; and those two serve in management roles. This is an imperfect mapping to our software roles, but using the acquisition career fields to refine the mapping has the advantage of allowing us to categorize the 690 personnel in our sample for whom software role is unknown.

As with the degree field–based analyses, we present the underlying salary-by-experience amounts for this occupation-based analysis in Figures B.5 and B.6 and the differential in Figure B.7. This analysis indicates that those most directly involved in the production of software within DoD may be paid less than their industry counterparts from the beginning of their careers, while those with oversight responsibilities may often earn a significantly higher salary in DoD. We are extremely hesitant, however, to say this is a finding or even an observation. The census groups are very broad, and our mapping to them is imperfect. Furthermore, our sample size is small enough that errors in that mapping may dramatically change the picture. Although it is true that we could have done a sensitivity analysis using different mappings, we were hesitant to devote additional resources to this inquiry. It may, however, be an area that researchers could pursue with a more complete DoD software workforce data set.

Table B.5
ACQ Career Field and Software Role Crosswalk to Census Occupations

ACQ Career Field	Software Role				
	Developer	Manager	SEIT	Unknown	Total
Business, cost estimating	0	0	0	1	1
Contracting	0	0	0	3	3
Engineering	662	46	228	563	1,499
Financial management	4	0	2	0	6
IT	5	0	13	5	23
N/A	0	1	0	4	5
Production, quality, and manufacturing	20	8	23	0	51
Program management	1	5	3	6	15
Science and technology	61	46	41	4	152
Systems engineer	7	2	1	1	11
Test and evaluation	1	0	2	103	106
Total	761	108	313	690	1,872
Comparison occupations:	Computer and information systems managers, engineering managers				
	Computer programmers and software developers, computer and information research scientists				
	Software quality assurance analysts and testers				

NOTE: "Engineering managers" is a subset of "architectural and engineering managers." We keep only those working in a subset of industries from this broad occupation. These industries are aircraft and parts manufacturing; aerospace products and parts manufacturing; computer systems design and related services; management, scientific, and technical consulting services; and scientific research and development services.

Figure B.5
Annual Salary Levels, by Experience/Age and Census Occupation Group (DoD Sample)

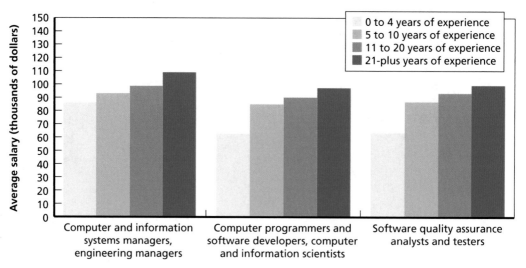

SOURCE: RAND analysis of collected DoD software personnel data and ACS one-year microdata from 2016 to 2018.

Figure B.6
Annual Salary Levels, by Experience/Age and Census Occupation Group (Industry Sample)

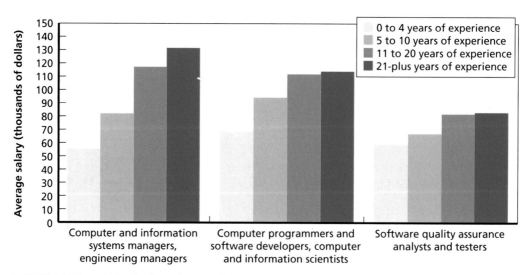

SOURCE: RAND analysis of collected DoD software personnel data and ACS one-year microdata from 2016 to 2018.

Figure B.7
Salary Profiles, by Experience and Census Occupation Group:
Salary Difference (DoD – Industry)

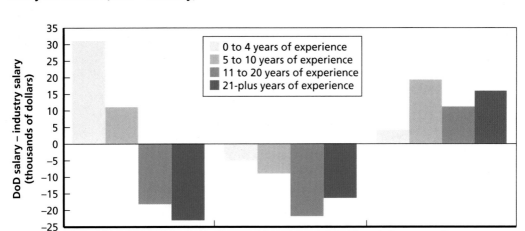

SOURCE: RAND analysis of collected DoD software personnel data and of U.S. industry workforce data derived from ACS microdata from 2016 to 2018.

Years of Experience Profiles by Pay Group

As noted earlier, our comparison of experience profiles between DoD and industry was marred by the fact that our industry data equated age with experience in software. A comparison of the resultant experience profiles revealed little except that the DoD profile was much more dynamic than that seen in the industry profile. In hopes of gaining additional insight into those dynamics, we attempted an analysis that would allow us to examine the DoD profile by *pay groups*. That is, we wanted to understand how the dynamics of years of software experience play out for more highly paid workers versus lower paid workers. Given the high correlations we observed earlier between pay and software experience, we felt this examination could reveal interesting dynamics unobscured by that correlation.

We generated this ex ante classification by comparing each alternate pay band with its GS scale equivalent (the GS scale is commonly used as a comparison metric for these alternate, broader pay bands that are part of the ACQ Demo project and related pay scales), then took the closest approximation in each pay scale to the groupings used in the NH pay scale. These groupings are illustrated in Figure B.8.

In Figures B.9 and B.10, we present the distribution of experience within our sample for each of these pay groups. Note that the sample only contains workers in pay groups 2 through 4. The results are first presented as a distribution of density (Figure B.9), similar to other analyses, then as a set of box plots (Figure B.10) showing the median value (the white line inside the box), the 25th and 75th percentile values (the upper and lower ends of the colored box), the "adjacent values"—the bar at

Figure B.8
Groupings of Multiple Pay Bands into Common Pay Groups

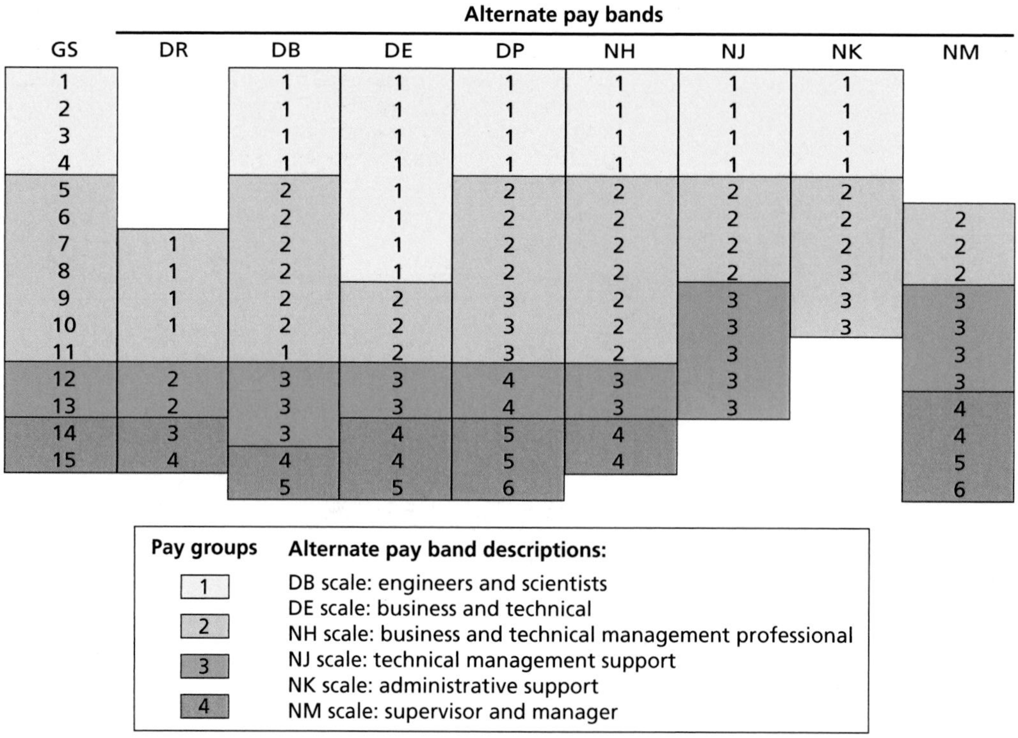

GS	DR	DB	DE	DP	NH	NJ	NK	NM
				Alternate pay bands				
1		1	1	1	1	1	1	
2		1	1	1	1	1	1	
3		1	1	1	1	1	1	
4		1	1	1	1	1	1	
5		2	1	2	2	2	2	
6		2	1	2	2	2	2	2
7	1	2	1	2	2	2	2	2
8	1	2	1	2	2	2	3	2
9	1	2	2	3	2	3	3	3
10	1	2	2	3	2	3	3	3
11	1	1	2	3	2	3		3
12	2	3	3	4	3	3		3
13	2	3	3	4	3	3		4
14	3	3	4	5	4			4
15	4	4	4	5	4			5
		5	5	6				6

Pay groups	Alternate pay band descriptions:
1	DB scale: engineers and scientists
	DE scale: business and technical
2	NH scale: business and technical management professional
	NJ scale: technical management support
3	NK scale: administrative support
4	NM scale: supervisor and manager

SOURCES: DoD, "Notice of Proposal to Design and Implement a Personnel Management Demonstration Project," *Federal Register*, Vol. 75, No. 163, August 24, 2010, p. 52147; DoD, 2019; U.S. Army Research Laboratory, "ARL Personnel Demo Pay Tables," January 2019.
NOTE: We generated these groupings ex ante and, ultimately, had no "pay group 1" personnel in our analysis sample. For categories DB5, DE5, DP6, and NM6, we assign average pay as 120 percent of GS15 pay, per the sources cited. For the small number (36) of active duty officers in the data we used the provided NH pay band equivalents; for a small number of cases (less than five) in which individuals identified as "SES," we used the same rule as for DB5/NM6 (GS15 × 120 percent).

the end of each "whisker," which indicates the most extreme values within the range given by the 75th percentile + 1.5 × (75th percentile – 25th percentile) and the analogous lower version of this same calculation—and any outlier values falling outside this range. In both plots, we can see evidence that there is a very compressed range of years of software experience in pay group 2, but there are many high outliers with experience ranging up to around 30 years. The experience of pay group 3 is primarily concentrated around ten years of experience, but there is a longer right tail of individuals with higher levels of experience (with another local peak around 33 years). Pay group 4 exhibits a strongly bimodal distribution of experience, with large numbers of individuals with experience clustered around 12 years and around 32 to 33 years. Additionally, the overall coverage of the experience of pay group 4 spans nearly the entire spectrum represented by the overall sample. This likely reflects the fact that senior pay grades

Figure B.9
Distribution of Years of Experience, by Pay Group for DoD Software Personnel (Kernel Density Plot)

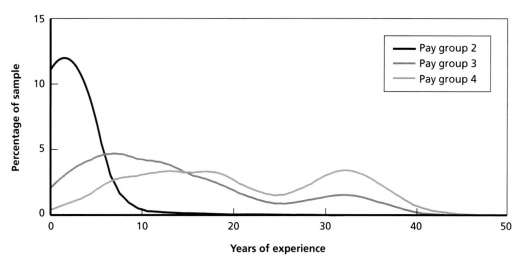

SOURCE: RAND analysis of DoD software workforce sample.

Figure B.10
Distribution of Years of Experience, by Pay Group for DoD Software Personnel (Box Plot)

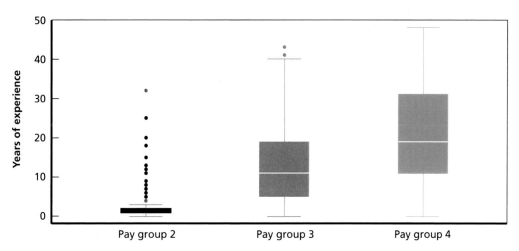

SOURCE: RAND analysis of DoD software workforce sample.

represent a combination of individuals with greater experience and individuals with specialized credentials or expertise, characteristics which do not necessarily imply the same sorts of experience levels. Table B.6 provides the mean and standard deviations for each pay group.

The vast majority of personnel with less than seven years of software experience fall into pay group 1: the lowest pay band. The smattering of personnel with less than seven years of software experience who fall into the higher pay bands may represent more-senior-level personnel who have only recently begun working in software, perhaps because of the increasing criticality of software to DoD programs. If so, this might be a positive trend. It also might be a sign that DoD is having to pay more to entice younger software professionals into the workforce. The mid-band of salary (pay group 3) is primarily made up of those with less than 20 years of software experience. The highest salary band (pay group 4) has a distinctly bimodal distribution: Although most individuals in this band have between ten and 20 years of software experience, another sizeable group is centered around 30 years of experience. In fact, at least half of software personnel with more than 25 years of software experience are also highly paid, suggesting these individuals may have a high level of competency.

Collecting the age of DoD software personnel could have clarified much of the dynamics of the workforce, as demonstrated by these analyses. Had we been able to collect our data in a way in which it could be merged with personnel records in the DMDC database as originally planned, we would know not only the age of the person but also the age at which they entered the DoD workforce and the age at which conversions from the uniformed or nonuniformed workforce to the industry workforce occurred. If future data fields are added to DMDC to track DoD's software workforce, we recommend DoD collect years of software experience in addition to whether an individual identifies as part of the workforce. Although the software competencies can be validated without collecting this information, both age and years of software experience are needed if we are to understand how DoD hiring, retirement, and retention policies are affecting those competencies.

Table B.6
Years of Experience, by Pay Group

Pay Group	Years of Experience		Observations
	Mean	Standard Deviation	
2	2.14	3.44	352
3	13.71	10.07	1,092
4	20.59	10.66	430
Total	13.12	11.09	1,874

SOURCE: RAND analysis of DoD software workforce sample.

Data Visualizations and Discussion of DoD Sample

In this appendix, we present additional visualizations of and discussions about the DoD sample, looking at characteristics by service branch, software role, and educational background. We hope these visualizations will allow those who so generously supplied data to us to better understand how their organizations fit within the range of software personnel we found.

Characteristics by Service Branch

Figure C.1 illustrates how software development roles are distributed by service organization. It is of note that the Air Force has the most equal distribution across the roles of developer, SEIT, and manager. This could be due to greater specialization (i.e., the Air Force has managers who only manage versus Navy and Army managers who manage

Figure C.1
DoD Software Personnel, by Role and Service Branch

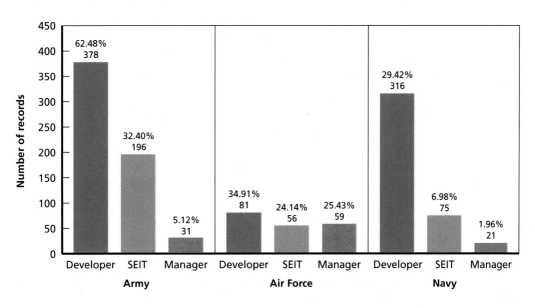

and perform development and SEIT roles), or it could be that the Air Force has a greater dependency on contractors for the development role. It may also be true that we simply do not have enough diversity in our Air Force sample for it to be representative of all Air Force software acquisitions.

Air Force organizations were unique within the sample for employing a uniformed and nonuniformed hybrid workforce. Within the three Air Force organizations surveyed, 8 percent of the workforce were uniformed personnel.[1] They ranged in rank from O-1 to O-6 and were evenly distributed among software developers, managers, and SEIT. More than 80 percent held an advanced degree of some kind, with aerospace and electrical engineering being the most common educational background.

As noted earlier in our discussion of software roles, Air Force organizations within the sample have noticeably different structures relative to their Army and Navy counterparts, with an almost one-to-one ratio between those who perform software development and those who manage software. This top-heavy structure translates into the largest percentage of NH-04 equivalent personnel for the Air Force at 54 percent, compared with 19 percent in the Navy and 4 percent in the Army (Figure C.2). The Air Force also has the largest percentage of personnel with doctoral degrees, constituting 26 percent of its workforce. This is compared to just 3 percent in the Army and 2 percent in the Navy (Figure C.3).

Each service branch had a different percentage of personnel focused on software-related tasks. Of those responding to the question, "Is software your primary duty?" 81 percent of Navy personnel said "Yes," compared with only 45 percent of Air Force software personnel in our sample. Army personnel were evenly split between each answer. We caution, however, that this statistic may not mean anything, given the difficulties that action officers had in responding to this question. The differences may be more a reflection of how the data were collected than of how the different services employ their software workforce.

Characteristics by Software Role

To better understand the distribution of DoD personnel within the software acquisition workflow, we characterized the software role of each individual. As described in Appendix B, individuals were grouped into developers, managers, and SEIT using data from their position description and software competencies. Unfortunately, approximately 37 percent of software roles could not be determined from the information we collected and thus are categorized as "Unknown." Missing data tended to cluster by organization, with certain organizations lacking information on software competencies, position description, or both.

---

[1] The three organizations were AFLCMC, Armament Directorate, Eglin; AFLCMC, Engineering Directorate, Avionics Division; and AFRL, Munitions Directorate.

Figure C.2
DoD Software Personnel Across Service Branches, by Rank (NH Equivalent)

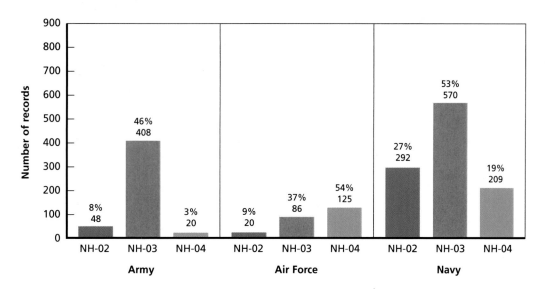

Figure C.3
DoD Software Personnel Across Service Branches, by Terminal Degree

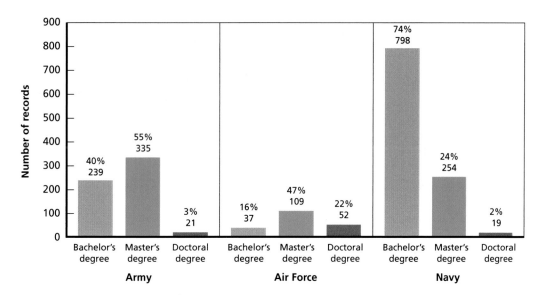

Of the individuals for whom we have data, the majority were developers (64 percent), followed by SEIT (27 percent) and managers (9 percent). Unsurprisingly, those in managerial roles received the highest compensations, with 52 percent at the NH-04 pay grade or its equivalent, compared with 28 percent of SEIT and just 8 percent of developers. Managers also had the highest levels of diversity in terms of educational background and career field.[2] For example, only 21 percent of managers had an educational background in software, compared with 49 percent within the total sample (Figure C.4). Similarly, although the "Engineering" acquisition career field made up

Figure C.4
DoD Software Personnel Across Software Roles, by Educational Background

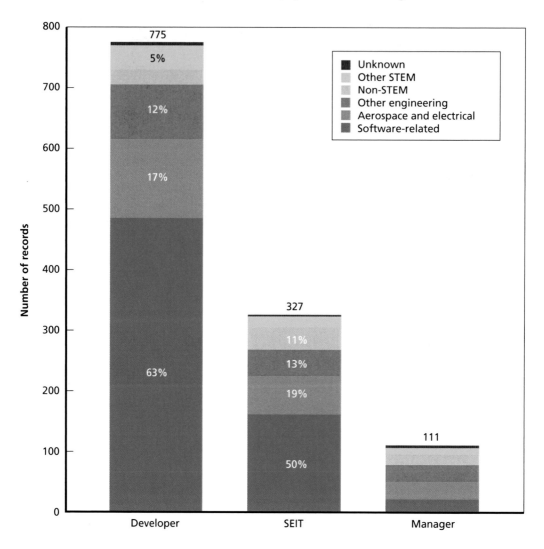

_____

[2]   Diversity is measured by the Simpson's Diversity Index.

80 percent of the total sample, engineers represented only 43 percent of managers (Figure C.5). Individuals in managerial and SEIT roles tended to attain high levels of education, with 62 percent of SEIT and 65 percent of managers holding an advanced degree, such as a master's or Ph.D., compared with only 42 percent of developers.

Characteristics by Educational Background

We analyzed the educational backgrounds of the sample to understand how those with software training were used across DoD. As indicated in Table B.3, approxi-

Figure C.5
DoD Software Personnel Across Software Roles, by Career Field

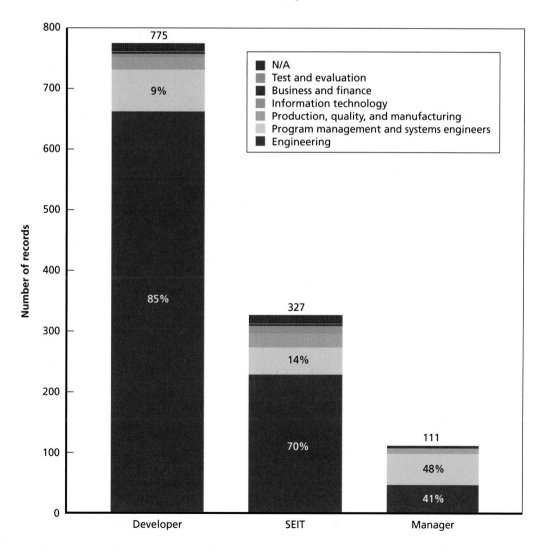

mately 49 percent of the sample had some type of educational background in software, and 31 percent had a terminal degree in aerospace or electrical engineering. Interestingly, those with educational backgrounds that diverged from their software acquisition role tended to have higher levels of education. For example, 22 percent of those with educational backgrounds in a STEM field other than engineering, as shown in Figure C.6, had a doctoral degree, compared with only 5 percent of those with an aerospace or electrical engineering background and 1 percent of those with a software-related background.

Although we caution against reading too much into the data regarding whether software is a primary duty, those with a software-related background were significantly more likely to have software as their primary duty compared with all other roles. As shown in Figure C.7, 80 percent of those with software-related backgrounds performed software as their primary duty, compared with just 45 percent of the rest of the sample.

Finally, each organization differed in terms of the distribution of educational backgrounds within their personnel, as shown in Figure C.8. A total of 93 percent of Navy personnel had educational backgrounds in software or aerospace and electrical engineering, compared with just 64 percent within the rest of the sample.

In addition to examining formal education data, we examined data regarding acquisition certification level required versus achieved. The level of acquisition certification required for personnel ranged from levels 1 to 3. Within our sample, 87 percent of individuals met or exceeded their required certification level, including 96 percent of individuals with a required level of 1 and 95 percent of individuals with a required level of 3. However, individuals with a required certification level of 2 met this requirement at a much lower rate (72 percent). Similarly, we found a statistically significant, positive relationship between job tenure and certification requirements met. Of those on the job for longer than the median tenure of ten years, 91 percent met or exceeded their certification requirements, compared with only 72 percent of newer personnel. In regard to software role, developers were the most likely to have incomplete certification requirements (86 percent had completed their certification requirements, which is to be expected according to the certification level and tenure of software developers).

Figure C.6
DoD Software Personnel, by Degree Type and Educational Background

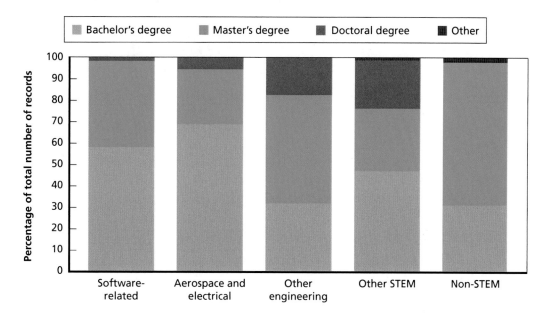

Figure C.7
DoD Software Personnel Who Perform Software as Their Primary Duty, by Educational Background

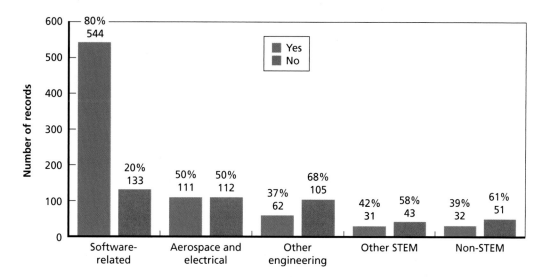

Figure C.8
DoD Software Personnel Educational Background, by Service Branch

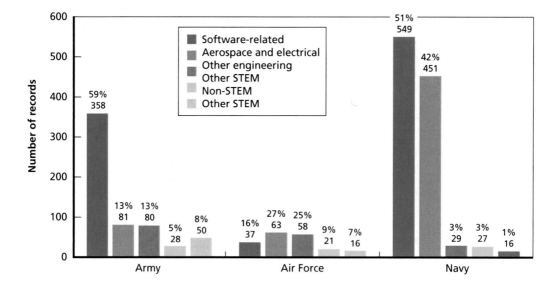

References

Beck, Kent, Mike Beedle, Arie van Bennekum, Alistair Cockburn, Ward Cunningham, Martin Fowler, James Grenning, Jim Highsmith, Andrew Hunt, Ron Jeffries, Jon Kern, Brian Marick, Robert C. Martin, Steve Mellor, Ken Schwaber, Jeff Sutherland, and Dave Thomas, "Manifesto for Agile Software Development," webpage, 2001. As of October 28, 2020: https://agilemanifesto.org

Bhise, Vivek D., *Automotive Product Development: A Systems Engineering Implementation*, Boca Raton, Fla.: CRC Press, 2017.

Brien, Spencer T., *Attrition Among the DoD Civilian Workforce*, Monterey, Calif.: Naval Postgraduate School, NPS-HR-20-004, October 29, 2019.

ComputerScience.org, "Exploring Software Engineering: A Comprehensive Guide to Careers and Top Employers," webpage, accessed June 2020. As of June 2020: https://www.computerscience.org/careers/software-engineer/

Defense Innovation Board, *Software Is Never Done: Refactoring the Acquisition Code for Competitive Advantage*, May 3, 2019. As of May 25, 2020: https://media.defense.gov/2019/May/01/2002126693/-1/1/0/SWAP%20MAIN%20REPORT.PDF

Department of Defense Directive 3000.09, *Autonomy in Weapon Systems*, Washington, D.C.: U.S. Department of Defense, Incorporating Change 1, May 8, 2017.

Department of Defense Instruction 1400.25, *Civilian Personnel Management*: Vol. 250, *Civilian Strategic Human Capital Planning (SHCP)*, Washington, D.C.: U.S. Department of Defense, June 7, 2016.

DoD—*See* U.S. Department of Defense.

Dutta, Raj Gautam, Xiaolong Guo, and Yier Jin, "Quantifying Trust in Autonomous System Under Uncertainties," *Proceedings*, 2016 29th IEEE International System-on-Chip Conference (SOCC), Seattle, Wash., September 6–9, 2016, pp. 362–336.

Gain, B. Cameron, "Microservices Security: Probably Not What You Think It Is," *New Stack*, March 26, 2018. As of June 10, 2020: https://thenewstack.io/microservices-security-probably-not-what-you-think-it-is/

Gansler, Jacques S., William Lucyshyn, and John Rigilano, *The Joint Tactical Radio System: Lessons Learned and the Way Forward*, College Park, Md.: University of Maryland Center for Public Policy and Private Enterprise, revised February 2012. As of June 10, 2020: https://apps.dtic.mil/dtic/tr/fulltext/u2/a623331.pdf

GAO—*See* Government Accountability Office.

Gates, Susan M., Edward G. Keating, Adria D. Jewell, Lindsay Daugherty, Bryan Tysinger, Albert A. Robbert, and Ralph Masi, *The Defense Acquisition Workforce: An Analysis of Personnel Trends Relevant to Policy, 1993–2006*, Santa Monica, Calif.: RAND Corporation, TR-572-OSD, 2008. As of June 22, 2020:
https://www.rand.org/pubs/technical_reports/TR572.html

Goldstein, Phil, "Which IT Skills Are Most in Demand in Federal IT?" *FedTech*, January 6, 2020. As of May 25, 2020:
https://fedtechmagazine.com/article/2020/01/which-it-skills-are-most-demand-federal-it

Government Accountability Office, *Defense Acquisition Process: Military Service Chiefs' Concerns Reflect Need to Better Define Requirements Before Programs Start*, Washington, D.C., GAO-15-469, June 11, 2015.

———, *IT Workforce: Key Practices Help Ensure Strong Integrated Program Teams; Selected Departments Need to Assess Skill Gaps*, Washington, D.C., GAO-17-8, November 2016.

———, *Weapon Systems Cybersecurity: DOD Just Beginning to Grapple with Scale of Vulnerabilities*, Washington, D.C., GAO-19-128, October 2018.

———, *DOD Space Acquisitions: Including Users Early and Often in Software Development Could Benefit Programs*, Washington, D.C., GAO-19-136, March 2019.

Gutgarts, Peter B., and Aaron Temin, "Security-Critical Versus Safety-Critical Software," *Proceedings*, 2010 IEEE International Conference on Technologies for Homeland Security (HST), Waltham, Mass., November 8–10, 2010, pp. 507–511.

Hodicky, Jan, "Autonomous Systems Operationalization Gaps Overcome by Modelling and Simulation," in Jan Hodicky, ed., *Modelling and Simulation for Autonomous Systems*, Third International Workshop, MESAS 2016, Rome, Italy, June 15–16, 2016, pp. 40–47.

Hough, Paul G., *Pitfalls in Calculating Cost Growth from Selected Acquisition Reports*, Santa Monica, Calif.: RAND Corporation, N-3136-AF, 1992. As of June 10, 2020:
https://www.rand.org/pubs/notes/N3136.html

Kelley, Patrick, "The Pentagon Is Facing a Serious Workforce Problem," *Government Technology*, July 10, 2019. As of May 2020:
https://www.govtech.com/workforce/The-Pentagon-Is-Facing-a-Serious-Workforce-Problem.html

Krazit, Tom, "How the U.S. Air Force Deployed Kubernetes and Istio on an F-16 in 45 Days," *New Stack*, December 24, 2019. As of June 10, 2020:
https://thenewstack.io/how-the-u-s-air-force-deployed-kubernetes-and-istio-on-an-f-16-in-45-days/

Leonard, Robert S., and Akilah Wallace, *Air Force Major Defense Acquisition Program Cost Growth Is Driven by Three Space Programs and the F-35A: Fiscal Year 2013 President's Budget Selected Acquisition Reports*, Santa Monica, Calif.: RAND Corporation, RR-477-AF, 2014. As of June 10, 2020:
https://www.rand.org/pubs/research_reports/RR477.html

Light, Thomas, Robert S. Leonard, Meagan L. Smith, Akilah Wallace, and Mark V. Arena, *Benchmarking Schedules for Major Defense Acquisition Programs*, Santa Monica, Calif.: RAND Corporation, RR-2144-AF, 2018. As of June 10, 2020:
https://www.rand.org/pubs/research_reports/RR2144.html

Long, James, "Army of Coders: Training the Force for the Multi-Domain Fight," Modern War Institute, December 21, 2018. As of June, 2020:
https://mwi.usma.edu/army-coders-training-force-multi-domain-fight/

Lyons, Joseph B., Matthew A. Clark, Alan R. Wagner, and Matthew J. Schuelke, "Certifiable Trust in Autonomous Systems: Making the Intractable Tangible," *AI Magazine*, Vol. 38, No. 3, Fall 2017, pp. 37–49.

Nahavandi, Saeid, "Trusted Autonomy Between Humans and Robots: Toward Human-on-the-Loop in Robotics and Autonomous Systems," *IEEE Systems, Man, and Cybernetics Magazine*, Vol. 3, No. 1, January 17, 2017, pp. 10–17.

National Institute of Standards and Technology, "Cyber-Physical Systems," webpage, undated. As of November 13, 2020:
https://www.nist.gov/el/cyber-physical-systems

Office of the Secretary of Defense, *Military Healthcare System (MHS) GENESIS Initial Operational Test and Evaluation (IOT&E) Report*, Washington, D.C., April 30, 2018.

Parsons, David, Teo Susnjak, and Anuradha Mathrani, "The Software Developer Cycle: Career Demographics and the Market Clock: Or, Is SQL the New COBOL?" *Proceedings of the ASWEC 2015 24th Australian Software Engineering Conference*, Vol. II, Adelaide, Australia, September 28–October 1, 2015, pp. 86–90.

Positive Technologies, *Web Application Vulnerabilities and Threats: Statistics for 2019*, February 13, 2020. As of June 10, 2020:
https://www.ptsecurity.com/upload/corporate/ww-en/analytics/web-vulnerabilities-2020-eng.pdf

Rierson, Leanna, *Developing Safety-Critical Software: A Practical Guide for Aviation Software and DO-178C Compliance*, Boca Raton, Fla.: CRC Press, 2013.

Robson, Sean, Bonnie L. Triezenberg, Samantha E. DiNicola, Lindsey Polley, John S. Davis II, and Maria C. Lytell, *Software Acquisition Workforce Initiative for the Department of Defense: Initial Competency Development and Preparation for Validation*, Santa Monica, Calif.: RAND Corporation, RR-3145-OSD, 2020. As of June 02, 2020:
https://www.rand.org/pubs/research_reports/RR3145.html

Schwartz, Moshe, and Charles V. O'Connor, *The Nunn-McCurdy Act: Background, Analysis, and Issues for Congress*, Washington, D.C.: Congressional Research Service, R41293, May 12, 2016.

Serbu, Jared, "Army Asks How Many Civilians, Contractors It Needs for Software Development in Weapons Systems," *Federal News Network*, October 10, 2016. As of June 10, 2020:
https://federalnewsnetwork.com/dod-reporters-notebook-jared-serbu/2016/10/army-asks-many-civilians-contractors-needs-software-development-weapons-systems/

Techopedia, "Embedded Software," webpage, undated. As of November 13, 2020:
https://www.techopedia.com/definition/29944/embedded-software

U.S. Army Research Laboratory, "ARL Personnel Demo Pay Tables," January 2019.

U.S. Bureau of Labor Statistics, "Occupational Outlook Handbook: Software Developers," webpage, accessed June 10, 2020. As of June 10, 2020:
https://www.bls.gov/ooh/computer-and-information-technology/software-developers.htm

U.S. Department of Defense, "Notice of Proposal to Design and Implement a Personnel Management Demonstration Project," *Federal Register*, Vol. 75, No. 163, August 24, 2010, pp. 52140–52171.

———, *Developmental Test and Evaluation: FY 2012 Annual Report*, Washington, D.C., March 2013.

———, *Defense Acquisition Workforce Program Desk Guide*, Washington, D.C., July 20, 2017a. As of March 14, 2018:
https://www.hci.mil/docs/Policy/Guidance%20Memoranda/DoDI_5000_66_Desk_Guide_Signed_20_July_2017.pdf

————, *Software System Safety: Implementation Process and Tasks Supporting MIL-STD-882E*, Washington, D.C., Revision A, October 17, 2017b.

————, "Major Defense Acquisition Programs (MDAP) and Major Automated Information Systems (MAIS) List," April 1, 2018.

————, *DoD Civilian Acquisition Workforce Personnel Demonstration Project (AcqDemo) Operating Guide*, Version 3, March 8, 2019.

U.S. Department of Defense Chief Information Officer, *DoD Enterprise DevSecOps Reference Design*, Washington, D.C., August 12, 2019.

Van Schooendenwoert, Nancy, and Brian Shoemaker, *Agile Methods for Safety-Critical Systems: A Primer Using Medical Device Examples*, Lean-Agile Partners and ShoeBar Associates, 2018.

Workscoop and Fedscoop, "Reskilling the Federal IT Workforce," undated. As of June 11, 2020: https://www.monstergovernmentsolutions.com/docs/federal-workforce-reskill-IT-demands-report.pdf